Introduction to Data Acquisition with LabVIEW

Dr. Robert H. King

Colorado School of Mines

Higher Education

Boston   Burr Ridge, IL   Dubuque, IA   New York   San Francisco   St. Louis
Bangkok   Bogotá   Caracas   Kuala Lumpur   Lisbon   London   Madrid   Mexico City
Milan   Montreal   New Delhi   Santiago   Seoul   Singapore   Sydney   Taipei   Toronto

INTRODUCTION TO DATA ACQUISITION WITH LABVIEW

Published by McGraw-Hill, a business unit of The McGraw-Hill Companies, Inc., 1221 Avenue of the Americas, New York, NY 10020. Copyright © 2009 by The McGraw-Hill Companies, Inc. All rights reserved. No part of this publication may be reproduced or distributed in any form or by any means, or stored in a database or retrieval system, without the prior written consent of The McGraw-Hill Companies, Inc., including, but not limited to, in any network or other electronic storage or transmission, or broadcast for distance learning.

Some ancillaries, including electronic and print components, may not be available to customers outside the United States.

This book is printed on acid-free paper.

1 2 3 4 5 6 7 8 9 0 DOC/DOC 0 9 8

ISBN 978–0–07–338584–6
MHID 0–07–338584–0

Global Publisher: *Raghothaman Srinivasan*
Senior Sponsoring Editor: *Bill Stenquist*
Director of Development: *Kristine Tibbetts*
Developmental Editor: *Darlene M. Schueller*
Senior Marketing Manager: *Curt Reynolds*
Senior Project Manager: *Kay J. Brimeyer*
Senior Production Supervisor: *Kara Kudronowicz*
Lead Media Project Manager: *Judi David*
Associate Design Coordinator: *Brenda A. Rolwes*
Cover Designer: *Studio Montage: St. Louis. Missouri*
(USE) Cover Image. *M-Series DAQ Board (left): DAQ Overview (center), and BNC-2120 Signal Accessory (right) (Photo Courtesy of Robert H. King): © Courtesy of National Instruments*
Senior Photo Research Coordinator: *Lori Hancock*
Compositor: *ICC Macmillan*
Typeface: *10/12 Times Roman*
Printer: *R.R. Donnelley Crawfordsville, IN*

LabVIEW, National Instruments, NI, and ni.com are trademarks of National Instruments Corporation

The credits section for this book begins on page 225 and is considered an extension of the copyright page.

Library of Congress Cataloging-in-Publication Data

King, Robert H.
  Introduction to data acquisition with LabVIEW/ Robert H. King.
    p. cm.
  Includes index.
  ISBN 978–0–07–338584–6 — ISBN 0–07–338584–0 (hard copy : alk. paper) 1. LabVIEW. 2.
Computer programming. 3. Computer graphics. 4. Laboratories–Computer programs. I. Title.

QA76.6.K5674 2009
006.6–dc22

                              2008023015

www.mhhe.com

Higher Education

INTRODUCTION TO DATA ACQUISITION WITH LABVIEW

Published by McGraw-Hill, a business unit of The McGraw-Hill Companies, Inc., 1221 Avenue of the Americas, New York, NY 10020. Copyright © 2009 by The McGraw-Hill Companies, Inc. All rights reserved. No part of this publication may be reproduced or distributed in any form or by any means, or stored in a database or retrieval system, without the prior written consent of The McGraw-Hill Companies, Inc., including, but not limited to, in any network or other electronic storage or transmission, or broadcast for distance learning.

Some ancillaries, including electronic and print components, may not be available to customers outside the United States.

This book is printed on acid-free paper.

1 2 3 4 5 6 7 8 9 0 DOC/DOC 0 9 8

ISBN 978–0–07–338584–6
MHID 0–07–338584–0

Global Publisher: *Raghothaman Srinivasan*
Senior Sponsoring Editor: *Bill Stenquist*
Director of Development: *Kristine Tibbetts*
Developmental Editor: *Darlene M. Schueller*
Senior Marketing Manager: *Curt Reynolds*
Senior Project Manager: *Kay J. Brimeyer*
Senior Production Supervisor: *Kara Kudronowicz*
Lead Media Project Manager: *Judi David*
Associate Design Coordinator: *Brenda A. Rolwes*
Cover Designer: *Studio Montage: St. Louis. Missouri*
(USE) Cover Image. *M-Series DAQ Board (left): DAQ Overview (center), and BNC-2120 Signal Accessory (right) (Photo Courtesy of Robert H. King): © Courtesy of National Instruments*
Senior Photo Research Coordinator: *Lori Hancock*
Compositor: *ICC Macmillan*
Typeface: *10/12 Times Roman*
Printer: *R.R. Donnelley Crawfordsville, IN*

LabVIEW, National Instruments, NI, and ni.com are trademarks of National Instruments Corporation

The credits section for this book begins on page 225 and is considered an extension of the copyright page.

Library of Congress Cataloging-in-Publication Data

King, Robert H.
  Introduction to data acquisition with LabVIEW/ Robert H. King.
    p. cm.
  Includes index.
  ISBN 978–0–07–338584–6 — ISBN 0–07–338584–0 (hard copy : alk. paper) 1. LabVIEW. 2.
Computer programming. 3. Computer graphics. 4. Laboratories–Computer programs. I. Title.

  QA76.6.K5674 2009
  006.6–dc22

                          2008023015

www.mhhe.com

Dedication

To Nancy, Ryan, and Alyssa

Chapter   5

DAQ State Machines   95

Chapter   6

Arrays   122

Chapter   7

Input and Output   159

Chapter   8

High-Frequency DAQ   192

Chapter   **9**

Summary   213

List of Figures

List of Tables

Introduction to Data Acquisition with LabVIEW teaches how to automatically measure physical properties with a computer data acquisition system or instruments. It uses numerous examples and the National Instruments (NI) LabVIEW graphical programming environment to lower the barriers to learning and to reduce the time required to successfully automate measurements. The text is designed for students in introductory college courses. It can also be used by experienced text-based language programmers to start learning graphical programming.

Scientists need data from measurements about the physical properties under study. Engineers need measurement values of physical properties as input to and verification of designs and numerical models and for product-quality testing. It is difficult to make enough high-precision measurements by hand to satisfy the needs of scientists and engineers; therefore, the process needs to be automated with computers, using the process called data acquisition (DAQ). DAQ is a very popular technology for industrial applications because it automatically provides essential information for product testing and new product design. DAQ improves speed and precision by controlling computer hardware with software. Low-level development of DAQ systems can take considerable time to master; therefore, the National Instruments (NI) Company has developed systems of hardware and software that significantly reduce the effort necessary to develop DAQ applications. When you combine this text with the NI system, you will quickly learn how to build automated measurement applications even though you have little or no experience with DAQ or programming.

The text was developed from materials used in the data acquisition course at the sophomore level in mechanical engineering at the Colorado School of Mines. Students learn DAQ in active sessions that begin with the instructor working an example application projected onto a screen in a computer laboratory where the students can work along with the instructor. The instructor explains the application thoroughly and answers student questions. The active-learning presentation is followed by a laboratory session in the same room where students work exercises in collaboration with other students, the instructors, and graduate teaching assistants. Both the active presentation and the exercises are included in this text.

The course at Colorado School of Mines concludes with an examination. The examination has been modified for use by NI to provide an international certification examination for students. The student certification in LabVIEW proves knowledge of DAQ with LabVIEW and consequently aids in gaining employment with NI partner companies.

Since the material is presented through examples, the text has also been used successfully in a one-credit-hour seminar course during an academic semester and for self-study by independent learners. The seminar course is divided in two parts. In the first part of the session, students present their solution to some of the exercises assigned the previous week and discuss alternate solutions. Then the instructor works one of the examples for the class, introducing new material for study during the upcoming week. If the computers are available, the students can work the example along with the instructor.

Introduction to Data Acquisition with LabVIEW can be used at institutions where it is difficult to devote an entire course to DAQ. The material in the text can become a module in a custom text for project and design courses. For example, an instructor might build a custom text from materials on teamwork, project planning, DAQ, feedback control, and so on, to support a project course.

The author evaluated other published materials before investing the time to develop new material. No materials were found that integrated software design and DAQ with LabVIEW in a concise, introductory format. Software design is extremely important to stop the spread of poorly programmed applications used in industry and research today. These poorly programmed applications run slow, have errors such as race conditions, use too many computer resources, and are difficult to read, scale up, and maintain. The solution is for students to learn proven structured programming architectures such as State Machines and Producer Consumer for data acquisition applications.

Correct, successful DAQ applications require knowledge of hardware and software. This text approaches DAQ from both hardware interfacing and software development. There are so many hardware options that all could not be implemented, so an inexpensive commonly used device, the NI BNC-2120 was adopted This doesn't mean the text is useless without this device. It is used as an example, and the techniques for hardware interfacing in the text can be applied to a wide range of devices. The text can also be used without hardware because the NI software environment has a hardware simulation capability that can be substituted for the BNC-2120 throughout the book.

Many students learn text-based programming—like some version of C, Basic, or JAVA—in high school or in the first year of postsecondary education. So they usually learn text programming before they learn DAQ. Consequently, this text builds on their text programming knowledge using the tried and proven educational pedagogy of learning a new subject by building on previous knowledge. This doesn't mean readers will not understand the text if they don't have a text programming background. The text can be easily used by first-time programmers. In fact, the LabVIEW graphical language is an excellent environment for learning how to program correctly. Novice programmers will find themselves able to accomplish a great deal in a short period of time, and it is more enjoyable than learning the syntax and format common to text programming.

The DAQ example applications in the book aren't confined to making measurements. They also include analyzing and presenting data and sharing it over networks. One of the great things about the LabVIEW software is that it is a general programming language. It is not confined to data acquisition and control. So your investment in learning LabVIEW for DAQ can be useful for many other applications.

Obtaining LabVIEW

This textbook uses LabVIEW version 8.5 for the Microsoft Windows 2000, NT, and XP operating systems. However, many of the concepts are applicable to other versions of LabVIEW and other operating systems such as Mac OS, UNIX, and LINUX. VIs developed in one operating system (OS) can usually be ported to other operating systems. The NI website (www.ni.com) provides details on OS conversions.

There are several ways to obtain LabVIEW.

1. This text contains the Student Edition of LabVIEW. The LabVIEW Student Edition Software Suite is a single DVD that includes LabVIEW for Windows XP/2000 as well as selected LabVIEW modules and toolkits. All LabVIEW Student Edition VIs work with the professional version of the LabVIEW Full Development System, and vice versa. However, code may not be backward compatible unless saved for a previous version. The LabVIEW Student Edition for Macintosh OS is included on CD for learning LabVIEW programming concepts. Although not covered in this text, the DAQmx Base driver is required to perform data acquisition using Macintosh OS. For more information and a free download visit ni.com/mac.

2. Additional copies of LabVIEW can be purchased online from National Instruments at www.ni.com or
   National Instruments
   11500 N Mopac Expressway
   Austin, TX 78759-3504
   888-280-7645
   Students, you can purchase additional copies of the LabVIEW Student Edition software at www.ni.com/labviewse.

3. There are several packages to choose among, so sometimes it is better to consult a local sales representative before purchasing on the Web, and you can also purchase through the sales representative.

4. Many universities and corporations have licenses. Inquire to a system administrator about adding the software to the appropriate computer.

5. Some corporate licenses support loading LabVIEW on off-site computers.

Book Contents

This text uses a series of examples to teach automated measurement with graphical software.

Chapter 1 introduces automated measurement applications in research, education, and industry. It provides an overview of measurement fundamentals and the data acquisition process. Additionally, it defines virtual instruments. It uses two simple examples to compare text-based with graphical programming.

Chapter 2 introduces the software for data acquisition. It introduces the LabVIEW software development environment using a simple example program that prints the value of π. It briefly explains how software interacts with the computer. It examines the data-flow paradigm that separates graphical programming from text programming and explains how to use the Execution Highlighting various LabVIEW tools for visualizing data flow and debugging.

Chapter 3 introduces Data Acquisition, including interfacing to plug-in boards, signal conditioning, use of the BNC-2120 signal accessory, driver and configuration software, and develops a program that acquires temperature data.

Chapter 4 describes the fundamentals of software design and flow control for DAQ applications, including algorithms, pseudo code, flowcharts, and GUI layout and design. It introduces selection with a Case Structure, online Context Help, repetition with a while loop, and timing.

Chapter 5 explains techniques for correctly building larger applications, including hierarchical and State Machine programming. An example DAQ application is developed with a State Machine that incorporates enumerated types, a type-defined enumerated control, shift registers, and sub VIs.

Chapter 6 develops algorithms that use arrays to manipulate large data sets. An example State Machine is developed that simulates climatic data using a for loop, a Formula Node, graphs and charts, and clusters. The chapter explains caveats with LabVIEW's ability to automatically coerce data types.

Chapter 7 explains the fundamental paradigm used by data acquisition systems to communicate with external resources such as files, Internet communications, data acquisition, and communications with external instruments. An example demonstrates opening a resource, reading or writing to it, closing it, and handling errors. Chapter 7 introduces common instrument busses such as the General Purpose Instrumentation Bus (GPIB). It shows examples of instrument drivers and explains how to use the instrument Assistant Express VI.

Chapter 8 extends the introductory data acquisition techniques to high-frequency applications, such as measuring sound. It compares software and hardware timing. Determining how fast to sample signals is an important part of the chapter. Chapter 8 presents transforms to convert the signal to the frequency domain. The fundamentals are applied in a music application.

Chapter 9 summarizes the text by having students build a DAQ State Machine to review designing and planning an application, designing and building a GUI, initializing and configuring resources, acquiring data, analyzing data, displaying information, saving to file, closing resources, and testing.

Disclaimer

set forth in the Software License Agreement accompanying the software, and except as expressly set forth in such Software License Agreement, no other warranties, either expressed or implied, are made with respect to the software. The entire liability of NI, its affiliates, and their suppliers for the software is set forth in the Software License Agreement. Your acceptance of the terms of the Software License Agreement is a prerequisite to your installation and use of the software. Without limiting the foregoing, NI does not warrant, guarantee, or make any representations regarding the use, or the results of the use, of the software in terms of correctness, accuracy, reliability, or otherwise and does not warrant that the operation of the software will be uninterrupted or error-free. To the maximum extent permitted by applicable law, in no event shall NI, its affiliates, their suppliers, The McGraw-Hill Companies, or any of their employees or representatives be liable for any damages, including any special, direct, indirect, incidental, exemplary, or consequential damages, expenses, lost profits, lost savings, business interruption, lost business information, or any other damages arising out of the use or inability to use the software, even if any of such parties has been advised of the possibility of such damages.

Website

Teaching and learning resources are available on the website that accompanies this text. You can access this site at www.mhhe.com/king.

Electronic Textbook Options

This text is offered through CourseSmart for both instructors and students. CourseSmart is an online browser where students can purchase access to this and other McGraw-Hill textbooks in a digital format. Through their browser, students can access the complete text online for one year at almost half the cost of a traditional text. Purchasing the eTextbook also allows students to take advantage of CourseSmart's Web tools for learning, which include full text search, notes and highlighting, and e-mail tools for sharing notes among classmates. To learn more about CourseSmart options, contact your sales representative or visit www.CourseSmart.com.

Acknowledgments

I want to thank the reviewers for their contribution to the content, organization, and quality of this book. The reviewers include:

Carlos Calderón Córdova, Ing. *Universidad Técnica Particular De Loja*

Horacio V. Estrada, *University of North Carolina-Charlotte*

Gary E. Ford, *University of California, Davis*

Burford Furman, *San José State University*

Israel Dario Carrión Granda, Ing. *Universidad Técnica Particular De Loja*

Hector Gutierrez, *Florida Institute of Technology*

Jean-Michel Maarek, *University of Southern California, University Park Campus*

Jose Raul Castro Mendieta, *Universidad Técnica Particular De Loja*

Robert W. Newcomb, *University of Maryland*

John Oldenburg, *California State University, Sacramento*
Hans-Dieter Seelig, *University of Colorado*
Adam Wax, *Duke University*
John D. Wellin, *Rochester Institute of Technology*

This book grew out of materials developed for the Colorado School of Mines mechanical engineering field session course, the Multidisciplinary Engineering Laboratory (MEL) Program, and the National Instruments Training Center at CSM. Dr. Joan Gosink, the former director of the CSM Engineering Division, provided important support for the development of all of these programs. The MEL Program received partial support from the U.S. Department of Education Fund for the Improvement of Postsecondary Education (FIPSE), the National Science Foundation, and the Parsons Foundation. Dr. Nigel Middleton initiated MEL, and Dr. Tom Grover, Dr. Terry Parker, and the author developed the experiments and educational materials. The author is indebted to the many students, graduate teaching assistants, and adjunct professors who have provided ideas for MEL and the field session over the years, especially Dr. Phillips Bradford who authored the audio measurement application included in this text. Many people at the National Instruments Company have provided support for the above programs, especially Paul Sweat, whose strong support for CSM led to establishing the NI Training Center in the CSM outreach program. Other NI personnel who provided support include Brad Armstrong and Armando Valim, who initiated the NI LabVIEW Academy that uses the text for examination preparation; Jim Cahow, Erik Luther, and David Corney, who provided valuable review of the material; and Ryan King, who provided valuable ideas and criticism of the example applications. I would like to acknowledge Bill Stenquist of McGraw-Hill, without whose support and guidance the book would not have been published.

About the Author

Dr. Robert King is a professor in the Engineering Division at the Colorado School of Mines. He has an extensive background teaching and using LabVIEW and is a certified LabVIEW Developer. He integrates LabVIEW throughout the curriculum at CSM through the Multidisciplinary Engineering Laboratory (MEL) Program and CSM Field Session for undergraduate students, research collaboration for graduate students, and the National Instruments Training Center (NITC) at CSM for outreach students. The MEL program uses LabVIEW and data acquisition in a sequence of sophomore, junior, and senior laboratories for multidisciplinary measurement and control for students in civil, mechanical, electrical, and environmental disciplines. The Field Session is a 40-week course for approximately 130 mechanical engineering students each year in DAQ with LabVIEW. Dr. King teaches LabVIEW Basics I and II, Intermediate I and II, Real-time, Compact RIO (Reconfigurable Input/Output), and Field-Programmable Gate Array (FPGA) courses in the NITC. Dr. King authored the CompactRIO course.

Dr. King has been at CSM since 1981. Prior to that he taught at Pennsylvania State University. He did his bachelor-level studies at the University of Utah and received his PhD from Penn State. He has taught over 20 different courses at CSM. His recent teaching experience includes the MEL courses, Computer Aided Engineering (applied finite element analysis), Introduction to Robotics, and Field Session. His current research includes using LabVIEW and PXI hardware to precisely measure forces for designing equipment for NASA's lunar outpost. Dr. King has used LabVIEW in several recent projects, including wind-turbine dynamometer testing, airborne beryllium continuous monitoring, fuel-cell development, materials property measurement, and remote monitoring of ground movement in construction excavations.

Introduction

OUTLINE

THE AUDIENCE FOR THIS BOOK

This book helps people learn how to make measurements automatically with a computer equipped with the proper data acquisition hardware and the National Instruments LabVIEW software.

The book is written for college students studying engineering and science. Measurements are important for students in a laboratory class to enhance understanding of physical phenomena or theory, to participate in a research project where measurements are used to add to our knowledge of some phenomenon, or to prepare for a professional career.

Computer based measurement is an essential tool for engineers and scientists. It can do things that humans find tedious and boring, such as repetitively measuring something continuously for long periods of time. For example, can you imagine having the job of measuring the voltage of batteries at a manufacturing plant, connecting a meter to each battery and writing the battery serial number and the voltage in a notebook repetitively every day? That would not only be a very boring task, but the nature of the task would make it very prone to error. Automating the measurement with a computer is a better approach.

Computer based measurement can also do very complex tasks. For example, we use them to solve large systems of matrix equations repetitively to calculate the stress in all of the members of a large suspension bridge. Consequently, most university engineering and science curricula require an introductory course in programming and developing software, and then later courses rely heavily on software in

1

engineering and science problem solutions. Being very conversant and skilled in computer usage is a key skill for today's scientists and engineers.

The best way to learn about computers and software is to use them to solve engineering and science problems. Consequently, this text is based on examples of how to automatically acquire data. The word *automatically* is key in the preceding sentence. We will automate data acquisition with a computer and software so we don't have to read values from a meter and tediously write them in a notebook or retype them. We will also be able to gather and store data faster with a computer than is possible manually.

To reduce cost, engineers and scientists use off-the-shelf computers that can be used for entertainment, network communications, word processing, etc. out of the box. These off-the-shelf computers won't perform our special tasks without new computer programs. Designing and writing computer programs can be tedious and time-consuming, so many engineers and scientists prefer not to write programs. This book attempts to reduce the barriers to writing programs through example-based instruction and use of the LabVIEW graphical or iconic programming environment. The material in the book is presented concisely, and it contains several references for readers interested in additional information.

Most students study programming in an introductory course taught in a text-based language such as C, C++, C#, FORTRAN, Java, or Visual Basic before learning how to make measurements automatically. Consequently, the initial chapters of this text provide examples that contrast LabVIEW and text programming to facilitate learning. This doesn't mean that students must know text-based programming before learning data acquisition. Readers without any prior knowledge of programming will also be able to learn data acquisition using this text.

INTRODUCTION TO MEASUREMENTS

Measurements give us values for variables. Some simple example measurements are when we weigh ourselves, look at the speedometer, or look at a clock. Weight, speed, and time are variables, and we want to know their values. We need a device to make the measurement, such as a scale, a speedometer, or a clock. We also need to have **confidence** that the value has minimal error. These are all local measurements, things in our presence. However, computer networks like the Internet allow us to make distributed measurements. For example, we can get the value of temperature in many places in the world over the Internet.

Engineers and scientists measure many things for many reasons. A few examples follow:

Input to a design—One of the author's projects is providing information to NASA for the design of equipment to develop the lunar outpost. An important input to equipment design is the force necessary to excavate regolith, so we are precisely measuring the force with the device shown in Figure 1.1 to provide this important design input value.

Evaluate a model—Engineers and scientists develop mathematical models to simulate physical systems. In addition to measuring the excavation force, we also use mathematical models to predict the force in different conditions because the regolith properties vary over the lunar surface. We compare measurement data with modeling results to validate our model.

Test a product before shipping—Companies test their products prior to shipping to assure quality and to comply with regulations. For example, Microsoft uses LabVIEW and NI PXI hardware to test the Xbox 360 wired and wireless controllers as shown in Figure 1.2.

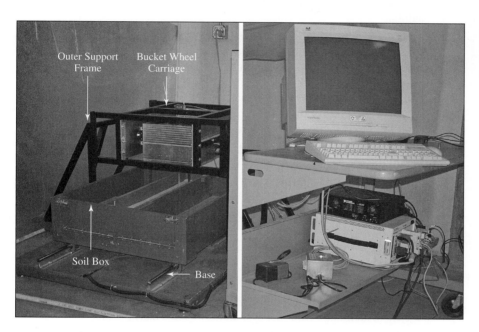

FIGURE 1.1
Soil-excavation-force measurement apparatus

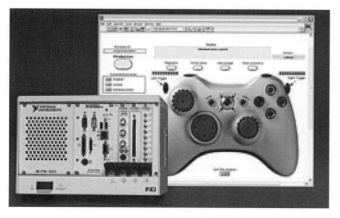

FIGURE 1.2
Microsoft Xbox 360 test system (ni.com)

Conserve energy—A programmable thermostat measures temperature and compares the value with one the user sets, controlling heating and cooling such that energy consumption is reduced, for example, by reducing heater output during times when no one is present.

Help a machine operator—Sometimes measurements help people control machines; for example, antiskid and antilock breaking systems provide additional safety for automobile drivers.

Understand and preserve the environment—Measurements are useful in evaluating and protecting the environment. We recently completed a project where we measured the temperature and humidity in several abandoned underground mines that endangered species of bats use for habitat.

Provide safe workplaces—Sometimes workers are exposed to hazards. We recently researched an instrument that would continuously measure small quantities of beryllium in the air at manufacturing facilities.

Engineers and scientists design measurement systems for applications like those described above. There are several texts on the subject of experimental design, and while this text will not replicate those publications, it is important to introduce the basic concepts to build an understanding of the motives for using data acquisition. As with all designs, we must gather as much information as possible about the measurement application. The following lists some desired information:

What variables should be measured?

What is the range of the values to be measured?

How precise should the data be?

Where will the measurements take place? Should we work over a network?

How often do we need to take data points?

How critical is timing? Does the system have to be deterministic?

Over what time duration do we need to measure?

How reliable does the system have to be? What is the cost of missing some data points?

What budget is available to make the measurements?

What instruments/transducers can be used?

Which are the best instruments/transducers for the budget, precision, environment, and so on?

How much data do we have to collect and manage?

What final result reports do we need to generate?

What types of signals are produced from the instruments/transducers?

What is the range of values produced from the instruments/transducers?

What noise will be present in the system?

What standards should be followed in making the measurement?

There are several errors possible in measurement systems. An error is the difference between the true value and the measured value. Because some error is present in most measurements, we use the terms *accuracy* and *precision* to reflect the amount of error present:

% accuracy = 100 * (error/true value)

% precision = 100 * (measurement − mean of n steady state measurements)/
(mean of n steady state measurements)

The following is a list of common transducer/instrument errors.

> Hysteresis—If you make a set of sequential measurements upscale and down-scale and there is a difference in the values, the error is called hysteresis.
> Linearity—Some transducers/instruments have a linear relationship between the property measured and their output value. Linearity error occurs when the measured value doesn't exactly fit the linear relationship.
> Repeatability—A repeatability error occurs when you measure the same value repeatedly over time and the value varies.
> Bias—Some transducers/instruments give consistently high or low values, causing a bias error.
> Resolution—Errors result from the analog to digital conversion process in data acquisition. They will be covered in detail in this text.
> Zero offset—If the transducer/instrument should read zero, but gives a nonzero value, it has a zero-offset error.
> Dynamic—Dynamic variables vary with time. An error occurs when the dynamic response of the transducer/instrument does not instantaneously capture the variable value at the time the measurement should occur.
> Overall—The overall error of a transducer/instrument is the square root of the sum of all the instrument errors.

Some errors are caused by the person making the measurements:

> Reading—reading the value incorrectly
> Dynamic reading—inability to read and record the data quickly enough to capture the dynamic variation in the values
> Interpolation—incorrect interpolation between markings on a meter
> Misusing an instrument—not following correct procedures
> Misapplication of an instrument—using the wrong transducer/instrument for the measurement
> Inadequate calibration—using an instrument without knowing its errors
> Recording—typing or writing the measurement value incorrectly

OVERVIEW OF THE DATA ACQUISITION PROCESS

Some measurements are made by manually connecting, operating, and reading instruments such as a voltmeter or oscilloscope, writing the numbers into a lab notebook, and typing the data into a spreadsheet program for analysis and graphing. Many modern

FIGURE 1.3

The data acquisition process to measure average wind velocity (after the LabVIEW Measurements Manual at www.ni.com)

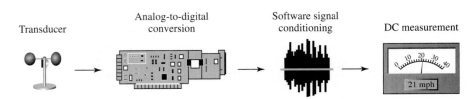

measurement systems use data acquisition (DAQ) to automate the traditional measurement process. The data acquisition process begins with transducers or instruments that convert physical properties, such as air velocity into electrical signals as shown in Figure 1.3. The electrical signals might be modified or conditioned, with devices such as amplifiers and filters, then, the signals are converted from analog to digital values that can be transferred to computer memory.

Table 1.1 lists some common transducers used in DAQ systems.

DAQ devices, such as the plug-in card shown in Figure 1.3, convert the signal to digital values that can be read by a computer and perform some signal conditioning. A DAQ device might reside in a slot of a desktop or a laptop computer or a personal digital assistant (PDA) or connect to the computer with a cable via a port such as the universal serial bus (USB) port.

TABLE 1.1   **Common Transducers for Measuring Physical Phenomena (after the LabVIEW Measurements Manual on ni.com)**

Phenomena	Transducer
Temperature	Thermocouples
	Resistance temperature detectors (RTDs)
	Thermistors
	Integrated circuit sensors
Light	Vacuum tube photosensors
	Photoconductive cells
Sound	Microphones
Force and pressure	Strain gages
	Piezoelectric transducers
	Load cells
Position (displacement)	Potentiometers
	Linear voltage differential transformers (LVDT)
	Optical encoders
Fluid flow	Head meters
	Rotational flowmeters
	Ultrasonic flowmeters
pH	pH electrodes

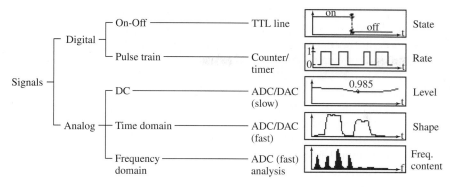

FIGURE 1.4
Common types of
signals (after the
LabVIEW
Measurements
Manual at ni.com)

Signals from transducers can be either analog or digital. A signal is classified as analog or digital by the way it conveys information. A digital (or binary) signal has only two possible discrete levels—high level (on) or low level (off). An analog signal, on the other hand, varies continuously with respect to time. An example is a sinusoid wave. Some typical signal types are shown in Figure 1.4 where TTL means transistor-to-transistor logic, ADC means analog-to-digital conversion, and DAC means digital-to-analog conversion.

Signals from transducers must be converted into a form a DAQ device can accept to precisely reflect changes in the physical property being measured. For example, the output voltage of most thermocouples is very small and susceptible to electrical noise. The noise might be larger than the actual change caused by temperature variation. Therefore, the transducer output might require amplification (a common form of signal conditioning) near the signal source to increase the signal to noise ratio. Other common types of signal conditioning include attenuation, filtering, linearization, transducer excitation, and isolation. Signal conditioning is the process of measuring and manipulating signals to improve accuracy, isolation, filtering, and so on. It is an important component of DAQ systems.

GRAPHICAL PROGRAMMING WITH LABVIEW

Before a computer will perform the repetitive tasks we require, we need to tell it, in detail, how to accomplish the task. We communicate the details of the task in a software program. This text helps students how to learn to write programs to measure physical properties, such as temperature and sound, for example. We can write programs that automatically control the process, retrieve the data from memory, display it, analyze it, and save it to files. It is much less time-consuming and more accurate to automatically enter the data into computer memory and save to a file than to manually write and type. We might read the voltmeter incorrectly or transpose numbers as we are writing, or we might make typing errors as we enter the data into a file. Avoiding the write and type steps reduces the time to complete an experiment. In addition, we might write the numbers into a lab notebook during the experiment and wait until later to graph the data. The DAQ and its software display the data graph on a computer monitor while we conduct the experiment, helping us identify mistakes before the experiment is completed. Furthermore, we automate a tedious task. If data are to be gathered over long periods, the computer will do it automatically, allowing us to do more interesting things. The computer can also do things we can't, such as record thousands of data points per second. Furthermore, some dynamic data change too quickly for humans to record, but computers are much faster and can measure very high-frequency signals. LabVIEW programs can also generate signals that automatically control instruments, motors, valves, and so on. So, the acquired data can be used in closed-loop control systems.

The programs we will write in LabVIEW interact with driver software (NI-DAQmx) to instruct the computer and the DAQ device how to perform the actual measurements and transfer them to the computer memory. Driver software is unique to the DAQ device or type of device. NI-DAQmx is the latest driver and includes the DAQ Assistant, which we will use to configure channels and measurement tasks. Our LabVIEW programs present and manipulate the raw data received through the DAQ device. Our programs also control the DAQ system by instructing the DAQ device when to acquire data, what transducer to use, how fast to acquire, how much to acquire, and when to stop.

We will use the LabVIEW graphical software development environment instead of traditional text programming in this text. We use LabVIEW because we can accomplish more in much less time. We program LabVIEW graphically with icons and virtual wires, instead of writing lines of text code with appropriate punctuation and indentation. Figures 1.5 and 1.6 compare graphical code in LabVIEW with text code in C that counts the number of zeros in an array of integers. There are several different ways to write this program in LabVIEW or with a text program. This program is not meant to be the most efficient code in either language, just a comparative example.

Figure 1.6 presents a comparable program written in the C text programming language.

The LabVIEW graphical program contains colored icons connected by colored virtual wires. The C program contains successive lines of text code, including symbols and punctuation.

The C program begins by defining and initializing the variable, count. LabVIEW accomplishes this by placing a rectangular icon, called an integer constant, in the program and typing 0 in it. If we were viewing the LabVIEW program in color, the

FIGURE 1.5 LabVIEW code that counts zeros in an array (from the LabVIEW Development Zone on ni.com)

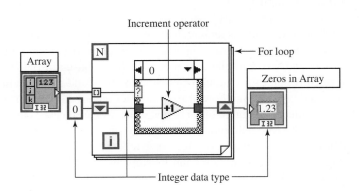

```
count=0;
if(array && (arraySize=GetArraySize(array)))
   for(i=0; i <arraySize; i++)
      if(array[i]==0)
         count ++;
printf("Located %ld zeros in array.",count);
```

integer constant and the wire connected to it would be blue. LabVIEW uses colors to indicate data types.

The computer executes C code from top to bottom, so next it executes the if(array &&(arraySize=GetArraySize(array))) line. If this line is true, it then executes the for(i=0; i <arraySize; i++) line. This line initiates a repetitive task (a for loop) that increments i by 1 until i reaches the array size. The next line evaluates the number in the array at position i to see if it is zero. If it is, the following line increments the count variable.

LabVIEW implements the repetitive task with an icon shaped like a stack of paper (a LabVIEW For Loop). The sequence of operation is not top down but is defined by the virtual data flow along the colored virtual wires leading from the array and count icons to the edge of the repetition structure. Another rectangular icon inside the LabVIEW For Loop contains the increment operator that increments count if the number at the array position is zero. When viewed in color, this icon is blue.

The C code printf("Located %ld zeros in array.",count); line executes last, writing a message and the number of zeros to the computer monitor. LabVIEW accomplishes the same task with the icon labeled "Zeros in array."

Because LabVIEW is a software development environment, it is much more than a programming language. It includes features that assist you to develop the program, debug it, make graphical user interfaces, communicate with external hardware, and distribute programs. We will discuss these in more detail throughout the text, especially communicating with external hardware.

LabVIEW can be used to build Virtual Instruments; for example, a virtual oscilloscope. In fact, LabVIEW is an acronym for Laboratory Virtual Instrument Engineering Workbench. A program developed with LabVIEW is usually called a Virtual Instrument, or VI, and the files have a .vi extension. Because Virtual Instruments are software based, an instrument for a specific application can be developed quickly and less expensively than a hardware-based stand-alone instrument. Traditional instruments also frequently lack portability, whereas virtual instruments run on portable platforms such as laptops, notebooks, or PDAs (personal digital assistants).

LabVIEW advancements have developed a general software development environment that has moved beyond its name. LabVIEW's success in its traditional focus—automated measurement and control applications—has supported its expansion to capabilities found in any program development environment: algorithms, user I/O (input/output), file I/O, printing, communicating over networks, and so on. It integrates these general software development features with automated measurement, control, and tools for developing graphical user interfaces (GUIs) like the one shown in Figure 1.7. Consequently, we can use LabVIEW for more than building

FIGURE 1.7

Graphical user
interface of the
vibration analysis
virtual instrument
(LabVIEW Examples)

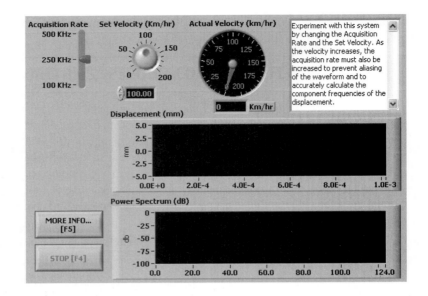

Virtual Instruments, because it is a general programming tool. For example, LabVIEW
programs solve linear algebra problems, evaluate calculus expressions, compute
statistics, fit equations to data, and more.

LabVIEW is a well-supported programming environment that is heavily used in
industry and academia. A literature survey provided more information on the popular-
ity of LabVIEW. A search of the Ei COMPENDEX found 389 publications in the
2000–2004 period that used LabVIEW as a keyword. The INSPEC database found
472 records in the same period. The authors represented over 30 countries from the
United States and over 50 U.S. universities. Publications are from fields such as bio-
medical, chemistry, civil engineering, electrical engineering, environmental science
and engineering, manufacturing, materials, nanotechnology, physics, security, and
robotics. Information on the history of LabVIEW is available at http://www.ni.com/.

SUMMARY

➢ The purpose of the book is to introduce data
acquisition and graphical programming.

➢ Many readers will find that they develop software faster in LabVIEW, and it will be more fun
because they can avoid the cumbersome strict
syntax and punctuation required with text
programming.

➢ LabVIEW developers work with icons and wires,
while text-language developers write lines of text
with appropriate punctuation and indentation.

➢ In addition to being a general programming
language, LabVIEW's initial intention was to
make computer-based virtual instruments that
are more flexible and portable than traditional
instruments.

➢ This book uses LabVIEW version 8.5 for the
Microsoft Windows 2000, NT, and XP operating
systems, but the material is directly applicable to
other versions of LabVIEW and other operating
systems.

EXERCISES

1. Create a handwritten notebook or a file in MS Word or Notepad with notes from the lecture/demonstration. Make a list of the topics covered and write a short description next to each topic. Keep this digital notebook up-to-date throughout the course. It will be checked at the end of each chapter.

2. Explain your programming background. What programming courses have you taken prior to reading this text? What software projects have you done? What programming languages have you used? How do you think graphical programming compares to your current programming practices? Write your answers in a word-processing file and save the file.

3. Science and engineering cover a large range of topics. List some areas of engineering or science that interest you: Some examples are automotive engineering, pollution control, material engineering or science, autonomous mobile robotics, and biomechanics.

   _____

   _____

   _____

   _____

   _____

4. Locate a piece of literature from the Web on an engineering or science project that interests you and write a brief summary of it in a word-processing file and save the file.

5. Is there an industry or academic application of automated measurement or control related to your area of interest?
   a. Locate a project description on the Web for automated measurement or control related to that area. Describe it in your notebook.
   b. Write the Web address (URL) in your notebook.
   c. Describe how computers are used in this project.
   d. Describe the international scope of the project.

6. Make a measurement and write all of the steps in fine detail in preparation for encoding the steps into a computer program. For example, measure the length of a computer keyboard in pencil lengths. Read your steps to a classmate and ask them to follow the steps to make the measurement. Tell your classmate they can't do anything unless told. They must pretend to be as dumb as a computer and will not do anything until you tell them. For example, you have to tell them to pick up the pencil. Revise the procedure if necessary until you have all the steps necessary for your simulated computer to make the measurement. Write the procedure in a word-processing file and save the file.

The LabVIEW Environment

INTRODUCTION

As explained in Chapter 1, automatically measuring physical quantities with a computer requires software and DAQ hardware. This chapter introduces the LabVIEW software development environment using a simple example program that prints the value of π. It briefly explains how software interacts with computer hardware and contrasts graphical and text programming. We will introduce DAQ hardware in Chapter 3.

OUTLINE

THE CIRCLE AREA PROGRAM

We are going to build the simple program shown in Figure 2.1, which is built quickly with the following steps:

> ➤ Start LabVIEW.
> ➤ Create a control on the Front Panel window for the radius.
> ➤ Create an indicator on the Front Panel window to display the area.
> ➤ Create 2 Multiply Functions on the Block Diagram window.
> ➤ Create a Pi constant on the Block Diagram window.
> ➤ Wire the Block Diagram objects.
> ➤ Save, run, and test the program.

We will discuss each step thoroughly in the following.

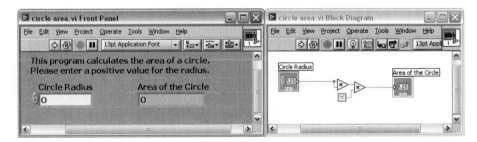

FIGURE 2.1
The circle area program

CONVENTIONS USED IN THE TEXT

This text presents programming concepts by using examples. The presentation uses the following set of conventions:

Select or **choose** means to move the mouse over an item and click (press) the left mouse button.

The double greater than sign, >>, means to follow a path from higher to lower levels. For example, the text in the next section includes the following path: Start >> Programs >> National Instruments >> LabVIEW 8.5. Move the cursor over Start (in the lower-left corner of the MS Windows Desktop), and press the left mouse button to pop up the start menu. Hover the cursor over the Programs item in the menu to reveal a sub menu of programs. Then hover the cursor over National Instruments to display another menu of the National Instruments programs, and select LabVIEW.

Ctrl-key means press the Control key and another key together, usually by holding the Control key down, pressing the additional key, then releasing both keys. For example, Ctrl-E. Many times these combinations represent keyboard shortcuts that can reduce program development time.

Double-click means to press the left mouse button twice in succession.

STARTING LABVIEW

Because the material presented in this text is based on examples, it is a good idea to launch LabVIEW and work along as you read the text, especially since the color of graphical objects is important and you can view the colors while programming in LabVIEW. To start LabVIEW, select Start >> Programs >> National Instruments >> LabVIEW X.XX where X.XX is the version number, 8.5, for example. If it is available, you can also double-click the LabVIEW icon on the desktop or in the quick-launch toolbar at the bottom of the screen. If you chose Start >> Programs >> National Instruments several programs are shown in addition to LabVIEW, such as DataSocket, VISA, Measurement Explorer, and so on. We will discuss some of these later. Please open LabVIEW and work the examples programs as you read.

When we launch LabVIEW, the Getting Started window shown in Figure 2.2 appears.

The Getting Started window allows the user to select from a set of files on the left and a set of resources on the right. The Resources column provides access to considerable information that can help a new user, including detailed descriptions of most palettes, menus, tools, VIs, functions, step-by-step instructions, links to the LabVIEW tutorials, PDF versions of all the LabVIEW manuals and application notes, and technical support resources. We will introduce these resources in more detail later but don't hesitate to explore any of the Getting Started window options. Similar information is available from the Help drop-down menu while programming.

FIGURE 2.2
The LabVIEW Getting Started window

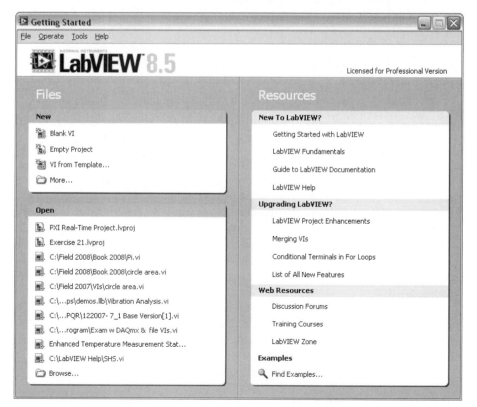

For our first program, we will start with a blank VI (virtual instrument). A blank VI is just a new program. We will discuss the other options for creating a new VI later but, again, don't hesitate to explore on your own.

Select Blank VI from the left column of the Getting Started window, and two windows similar to Figure 2.3 are displayed. The window in front, called the front panel window, is the graphical user interface (GUI). The window in back is the block diagram window, where we will develop the code for our programs.

Menu bar

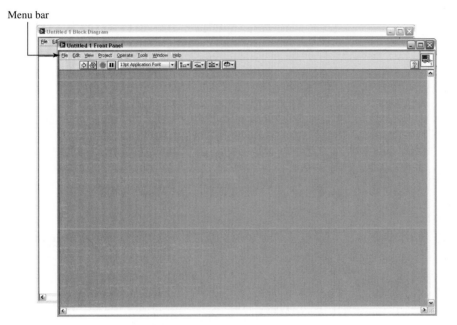

FIGURE 2.3
Blank VI

Traditional introductory computer-programming texts start with a simple program that prints something like "Hello World" to the monitor. Dietel and Dietel (2003) prints "Welcome to C++." Ingber and Etter (2003) compute and display the distance between two points, and Holloway (2004) calculates and displays the square of a number. Austin and Chancogne (1999) print the value of π using the C language. We can do any of these as our first LabVIEW program, and we will start by displaying the value of π.

DEVELOPING THE USER INTERFACE ON THE FRONT PANEL

The LabVIEW environment offers a large variety of features, but we don't have to learn all of them at once. We will learn just what we need to display π, and we will explore other functionality later as we need it.

We will build the GUI by using a set of tools to insert and manipulate objects on the front panel window. LabVIEW uses two fundamental objects called controls and indicators to input and output information. Our first program will output the value of π without requiring any input from the user, so we will use only an indicator. To place an indicator on the front panel window, move the cursor onto the front panel window and right-click to display the Controls palette (Figure 2.4). We can also

FIGURE 2.4 Controls palette

Thumbtack ⟶

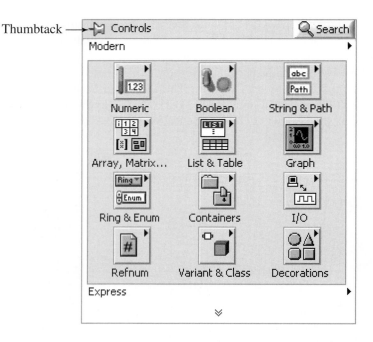

FIGURE 2.5 Numeric button on tacked-down Controls palette

Modern style ⟶

Numeric button ⟶

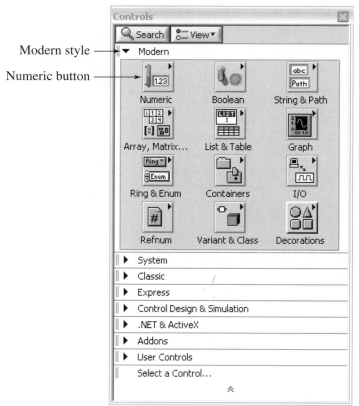

display the Controls palette from the View pull-down in the menu bar (View >> Controls Palette).

Idle the cursor over the thumbtack (Figure 2.4) in the upper-left corner of the palette. Notice that the thumbtack moves into the hole. Left-click when the thumbtack is in the hole. This will tack the palette down and display the options shown in Figure 2.5.

The Controls palette contains a large number of objects, arranged in categories in a tree structure. We will use a numeric indicator from the tree to display π. Select the Numeric category button to show the types of numeric objects available for use. Note that we will use a Numeric control from the Modern Controls palette. System and Classic palettes also have Numeric Controls.

As we move the cursor over the various objects in the palette, they are highlighted and identified by a tip strip, as shown in Figure 2.6. Select the highlighted Numeric Indicator, shown in Figure 2.6.

When we choose it, the Numeric Indicator icon attaches to the cursor, and the cursor shape automatically changes to a hand. Move the cursor to the desired location on the front panel window, as shown in Figure 2.7, and left-click again.

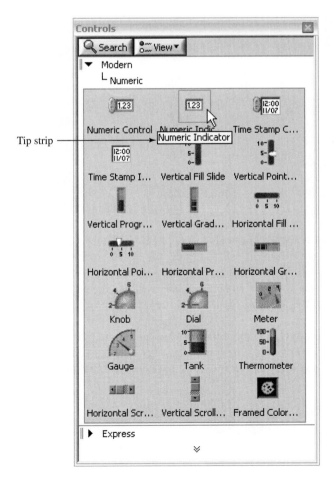

Tip strip

FIGURE 2.6 Using the tip strip and highlighting to select the Numeric Indicator

FIGURE 2.7 Placing a Numeric Indicator on the front panel window

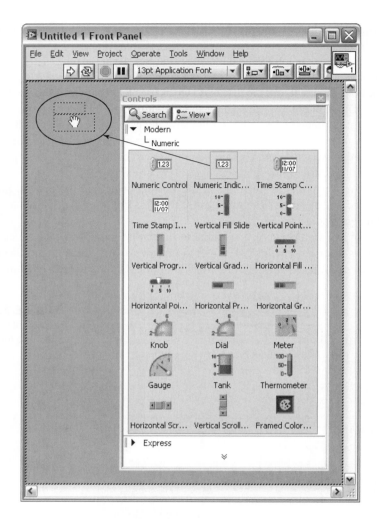

The title, or label, has a black background as soon as the indicator is placed on the Front Panel window, as shown on the left in Figure 2.8. When we first insert an object, its label background will be black, indicating we can change the title. Before we make any additional moves or clicks with the mouse, we can change the title of the control easily by typing a new name. So, type the new text, Pi, in the label, as shown on the right in Figure 2.8.

If we click the cursor before typing and lose the black background, edit the label by double-clicking on it and then begin typing the new label. The background turns black and cursor shape automatically changes to a vertical line, the Labeling tool, for editing or typing text.

FIGURE 2.8 Change the Numeric label to a more meaningful and specific word or phrase

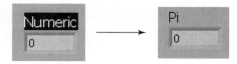

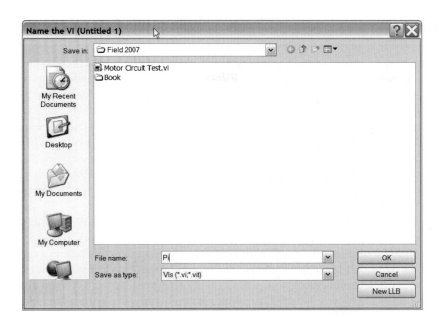

FIGURE 2.9
Saving a VI

Placing the indicator on the front panel window is similar to declaring a variable in text-based programming. When we execute the program, LabVIEW will communicate with the operating system to reserve some computer memory for the numbers in the Pi indicator and to display those numbers on the computer monitor.

In text programming, we usually have to define the data type, so the appropriate type of memory will be reserved. The numeric indicator is double-precision floating-point data type by default, so we don't have to define the data type unless we want to change to a different type. Choose File >> Save As and save the program as pi.vi in an appropriate folder, as shown in Figure 2.9. LabVIEW will automatically add the .vi extension. It is a good practice to save often while programming. After the first save, which names the file, use Ctrl-S to save frequently while working. Notice that after the program is saved, the title changes in the title bar. The title was Untitled 1 in Figure 2.7, and it changed to Pi.vi after saving (Figure 2.10). After the first save, which names the file, use Ctrl-S to save frequently while working.

Title bar —

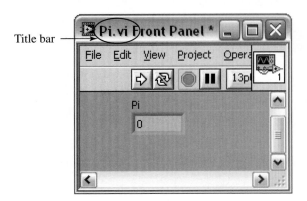

FIGURE 2.10 Pi.vi front
panel window

THE LABVIEW CURSOR AND TOOLS PALETTE

Because we are programming graphically, and not typing lines of text code, the cursor is a very powerful tool. The LabVIEW mouse cursor automatically changes shape to perform different operations, as you probably noticed in the preceding operations. When the Getting Started Window opened, the cursor changed to a hand icon when we selected a blank VI. After the blank VI opened, the cursor became a "plus" sign, or a crosshair icon. After we right-clicked to display the controls palette, the cursor changed to an arrowhead. Then it changed to the hand icon when placing the numeric control at the desired position on the front panel window. If you didn't notice all of these changes, run through the sequence again and watch the cursor morph between icons.

If automatic tool selection is enabled (the default), LabVIEW automatically changes the tool icon based on its position. We can disable automatic tool selection and select a tool manually by clicking the desired tool on the tools palette available from the View pull-down in the menu bar (View >> Tools Palette), as shown in Figure 2.11. Click the automatic tool selection button or press the <Shift-Tab> keys to enable automatic tool selection again. I recommend that you use the automatic tool selection feature. Only disable automatic tool selection if you are having difficulty getting the automatic tool to change to the shape you want.

Figure 2.12 explains the function of the various Tools Palette Icons.

Automatic tool switching requires that we **position the cursor very carefully**. For example, Figure 2.13 shows different components of the Pi indicator. As we move the cursor over each of these areas, the cursor shape changes. The cursor is an arrow shape when it is over either the control border or the label box border. It adopts the text shape when it is over the label text box or the indicator value box. If we click on the indicator border, we select both the indicator and the text label. But if we position the cursor over the label border, we select only the label. The borders are very narrow, and if we move the cursor slightly, it leaves the border area. This might happen accidentally if we slightly scoot the cursor unintentionally when we push a mouse button. So, hold the cursor steady when pushing the mouse buttons.

Users familiar with earlier versions of LabVIEW sometimes disable automatic mode and use the tools palette to select tools manually. Pressing the <Tab> key will

FIGURE 2.11 LabVIEW
Tools Palette

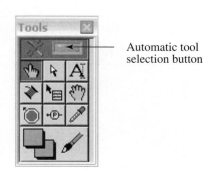

Automatic tool
selection button

Tools Palette Icons

FIGURE 2.12
Tools Palette Icons

The Tools palette contains the following tools, which you use to operate and modify front panel and block diagram objects.

 Automatic Tool Selection—If automatic tool selection is enabled and you move the cursor over objects on the front panel or block diagram, LabVIEW automatically selects the corresponding tool from the **Tools** palette. You can disable automatic tool selection and select a tool manually.

Operating—Changes the value of a control.

Positioning—Positions, resizes, and selects objects.

 Labeling—Creates free labels and captions, edits existing labels and captions, or selects the text within a control.

Wiring—Wires objects together on the block diagram.

Object Shortcut Menu—Opens the shortcut menu of an object.

Scrolling—Scrolls the window without using the scroll bars.

Breakpoint—Sets breakpoints on VIs, functions, nodes, wires, and structures to pause execution at that location.

Probe—Creates probes on wires. Use the Probe tool to check intermediate values in a VI that produces questionable or unexpected results.

Color Copying—Copies colors for pasting with the Coloring tool.

Coloring—Sets the foreground and background colors.

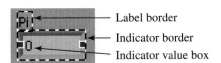
— Label border
— Indicator border
— Indicator value box

FIGURE 2.13 LabVIEW indicator components

cycle through the four most common tools, and pressing the space bar will switch to the next most useful tool.

This completes the front panel window of this simple VI.

DEVELOP CODE ON THE BLOCK DIAGRAM

To develop the code that tells the computer to display the value of π, we will place icons and wires on the block diagram window that was tiled behind the front panel window (Figure 2.3). Normally it isn't necessary to view the block diagram window when running the program. Therefore, we use the front panel window (GUI) to run the VI, input information, and view results. We use the block diagram window to develop, debug, and modify the program.

There are several ways to move the block diagram window to the forefront:

1. Click on its partially exposed edge in the tiled configuration.
2. Choose "Show Block Diagram" from the Window drop-down item on the menu bar.

FIGURE 2.14
Pi program front
panel and block
diagram windows

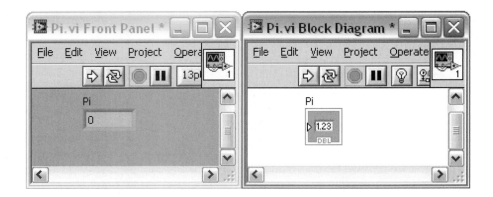

3. Use the Ctrl-E shortcut to toggle between the front panel and block diagram windows.
4. Use Ctrl-T to tile the front panel and block diagram windows left and right.

Ctrl-E and Ctrl-T are shortcut keys. They save a lot of time when programming. You might want to make a list and update it as new shortcut keys are introduced.

We will use Ctrl-T to observe the front panel and block diagram windows simultaneously because our program is small. If we resize the tiled windows, they will appear as shown in Figure 2.14. When we develop larger programs, we will use Ctrl-E.

Figure 2.14 shows that LabVIEW automatically added an icon to the block diagram window when we placed the numeric indicator on the front panel window and the block diagram icon has the same label. To observe this, delete the indicator icon and replace it while the windows are tiled. Use the undo button or the Ctrl-Z shortcut to delete the indicator icon.

The Pi terminal icon on the block diagram window has an arrowhead on the left pointing into the icon, shown in Figure 2.15. This identifies it as an indicator that will accept a data input. Consequently, we will send data from a constant representing the value of π to this terminal. If you are working in LabVIEW while reading this text, you will notice that the icon border color is orange, which identifies the double-precision floating-point data type as does the DBL text at the bottom of the icon.

We access block diagram objects, similar to front panel objects, with a palette. The palette used for block diagram objects is called the Functions palette. Right-click in the block diagram window to display it (Figure 2.16).

To find the π constant in the Functions Palette, click the Search button in the functions palette menu bar, and type Pi into the search field, as shown in Figure 2.17. Then, scroll down the list of results to select either of the two locations of the Pi constant.

Now that we found it, we will place the constant on the block diagram window. Double-click on one of the Pi constants in the search list and LabVIEW

FIGURE 2.15
Pi indicator icon

Arrowhead

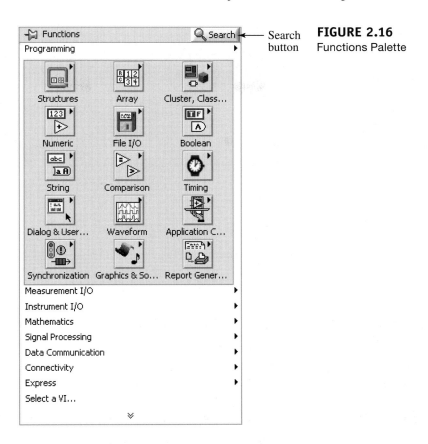

FIGURE 2.16
Functions Palette

Search button — Search

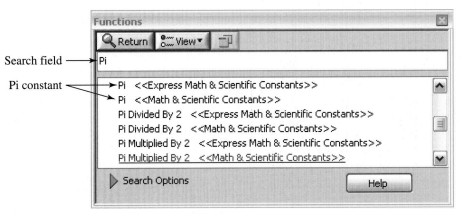

FIGURE 2.17
Using Search functions to locate the Pi constant

Search field — Pi

Pi constant — Pi <<Express Math & Scientific Constants>>
Pi <<Math & Scientific Constants>>

highlights the constant in its palette. Click the constant and it will attach to the cursor hand icon. Place it on the block diagram window to the left of the Pi indicator terminal, as shown in Figure 2.18. The functions palette shows the hierarchy to the owning palette (Math & Scientific Constants) in the palette tree just above the constant icons.

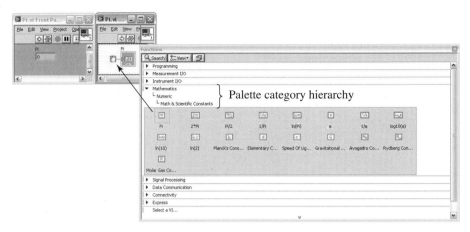

FIGURE 2.18 Placing the Pi constant on the block diagram window

When we execute the program, LabVIEW tells the computer to allocate memory for the double-precision floating-point π value and to place the value of π in memory. (This is similar to defining a constant in text-based programming.) In order to display that value on the computer monitor, we need to tell the computer to send the value to the indicator.

We tell the computer to transfer data in LabVIEW by stretching a wire between two icons. The first icon will execute and transfer data to the second icon. This is called data flow, and we will discuss it in more detail later. We use wires in LabVIEW like we would use physical wires between two hardware connectors when wiring an electrical circuit. We connect one end of the wire to a terminal on one icon and extend the wire to another icon where we connect the other end of the wire.

As we move or place an icon close to another icon on the block diagram window, LabVIEW will automatically construct the wire. As we move a selected object close to other objects on the block diagram, LabVIEW draws temporary wires to show valid connections. When we release the mouse button to place the object on the block diagram, LabVIEW automatically connects the wires. By default, automatic wiring is enabled when we select an object from the functions palette or when we copy an object already on the block diagram window. Automatic wiring is disabled by default when we use the Positioning tool to move an object already on the block diagram window. Toggle automatic wiring by pressing the space bar while you drag an object onto the block diagram or move an object using the Positioning tool. When automatic wiring is enabled, the selected object retains its appearance when you drag it. When automatic wiring is disabled, the selected object appears as a dotted outline when you drag it.

You can also construct the wire manually. To start the wire, move the cursor over the output terminal on the right side of the constant icon, as shown in Figure 2.19. The output terminal on the right of the constant icon blinks, and the cursor icon automatically changes to the wiring tool icon, which is shaped like a spool of wire. The cursor point of the tool is the tip of the unwound wire spool. Attach the wire to the terminal by left-clicking on the right side of the constant. Then, extend the wire to

Blinking output
terminal

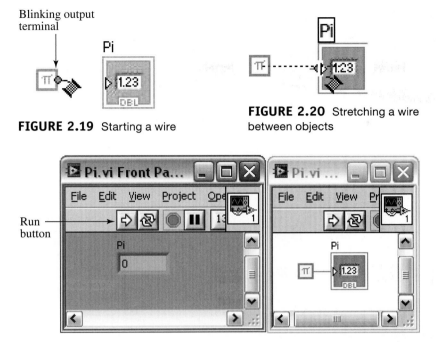

FIGURE 2.19 Starting a wire

FIGURE 2.20 Stretching a wire
between objects

Run
button

FIGURE 2.21 The pi program

the indicator by moving the cursor to the input terminal on the left side of the
Pi-indicator icon, as shown in Figure 2.20. Do not press the mouse button while moving the cursor. Until connected to the indicator, the wire is a dashed line. Connect the
wire when the Pi indicator terminal blinks by left-clicking again when the cursor is
over the indicator terminal. Be careful to hold the mouse steady when clicking. The
wire turns to a solid line after it is connected, as shown in Figure 2.21.

If you are working in LabVIEW while reading this text, you will notice that the
Pi constant, the wire, and the Pi indicator icon are all orange color showing that the
double-precision floating-point data type is used consistently through the sequence
of operations in this VI.

This completes the program, so save it again. To execute the VI, click the Run button, which is the white arrow on the left side of the Front Panel toolbar (Figure 2.21).
After execution, the Pi indicator value changes from 0 to 3.14159. We have now completed and tested our first LabVIEW graphical program.

DOCUMENTATION

In LabVIEW, it is possible to add free labels to the program and also document the
program, but the user interface and graphical nature of the program reduce the need
for comments. To add documentation, use File >> VI properties >> Documentation,
as shown in Figure 2.22.

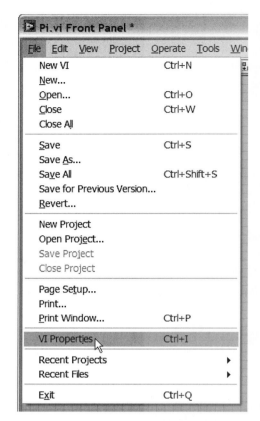

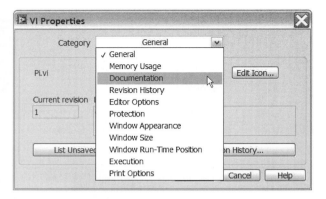

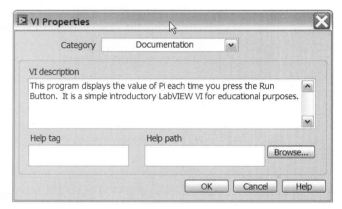

FIGURE 2.22 Adding documentation to a VI

Double-click anywhere on the front panel or block diagram windows to add a free label to provide additional documentation. Right-click controls or indicators and choose Properties to add documentation or tip strips specific to that front panel object. Select File >> Print to print VI documentation.

COMPARING LABVIEW WITH A C TEXT-BASED PROGRAM

We were able to develop the Pi VI without any previous knowledge of computer programming. Nevertheless, some readers may have experience with text-based programming and are curious about the transition to LabVIEW. We can compare the Pi graphical program with the C text-based program shown in Figure 2.23.

The C program contains comments, preprocessor directives, header files, a main function, and standard library functions.

The comments begin with /* and end with */. C program comments include information such as the purpose, author, and date. Sometimes C programmers use /* for a single line comment. We use labels, free labels, icons, wires, color, tip strips, and documentation instead of comments.

```
/*
 *
 * ====================================
 * Print value of "pi" to computer screen
 *
 * Written by: Mark Austin           January, 1994
 *====================================
 */

#include <stdio.h> /*Standard Input/Output function declarations*/

#define PI 3.1415926

int main (void) {

        printf("Approximate value of PI = %f \n", PI);
        return (0);
}
```

FIGURE 2.23 C text program that prints the value of pi to the computer monitor
Copyright © 1999 Austin. Reprinted with permission of John Wiley & Sons, Inc.

The #include <stdio.h> line in the C program directs the preprocessor to add the lines from a header file called stdio.h to the program. Stdio.h is the standard library for input and output that supports the printf statement used later in the program, which directs the output to the computer monitor. Header files and preprocessor directives are not necessary in LabVIEW. When you add the indicator to the front panel window, LabVIEW will automatically accomplish the same things as the header file.

The #define command reserves a space in memory for PI and assigns a value to it. We accomplished a similar task in LabVIEW when we placed the π constant on the block diagram window.

C variable names must be valid identifiers. An identifier is a name (a string of characters) that includes letters, digits, and underscores, but it cannot start with a number. C programmers use capitalization to differentiate between identifiers, depending on whether the first letter is capital or lowercase or whether all letters are capitals. Spaces are not allowed in multiple-word identifiers. You do not have to follow these rules when typing a label in LabVIEW.

All C programs are functions. The Pi program uses only the "main" function preceded by "int" for integer. The word "void" that is surrounded by parentheses tells the computer that this function does not have any input arguments. LabVIEW automatically takes care of this activity.

A brace punctuation symbol follows the main function statement. LabVIEW doesn't require these, or any, punctuation symbols.

The C program printf statement includes some formatting instructions, %f and \n, for floating-point formatting and a line feed. Printf defaults to six decimals of precision. The Pi indicator on the program front panel window accomplishes a similar

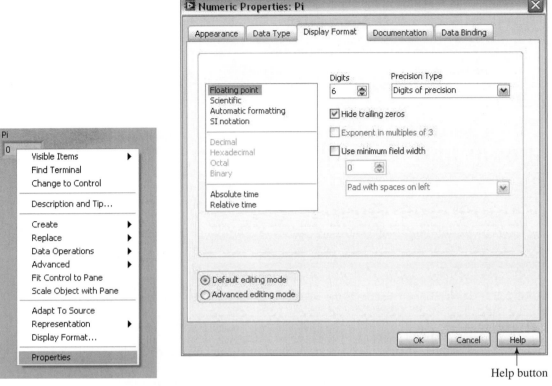

Help button

FIGURE 2.24 Pi indicator properties dialog box

task. The default data type for the numeric indicator that we used is a double-precision floating point, which is appropriate for displaying Pi and most other values. We did not have to give print format instructions (%f \n) in LabVIEW. The LabVIEW indicator defaults to six significant digits.

We can change the default significant digits and other properties of the indicator by right-clicking on the indicator to reveal the shortcut menu and selecting the last item, Properties (Figure 2.24). In the properties dialog box, choose the "Display Format" tab and change to digits of precision using the arrow on the right of the field, which allows you to select either significant digits or digits of precision. Note that you can change many more properties by choosing other tabs. Click the Help button on the lower right corner of the Properties window to learn more about the various options.

The printf statement in the C program is indented for readability. We align objects in LabVIEW so our wires are straight and to make our code readable. We aligned the π constant with the indicator terminal on the block diagram window so we had a straight wire, and we placed the constant close to the terminal to reduce the wire length.

The printf statement in C is followed by a semicolon. LabVIEW doesn't require punctuation to identify the end of statements.

Some common programming errors that might be made in the C program but are not problematic when using LabVIEW include the following:

> Forgetting the /* at the beginning or */ at the end of a comment or getting them reversed
> Typing print instead of printf
> Using a capital letter instead of a lowercase letter
> Using a variable before defining it
> Omitting a semicolon

After the C program is written, it is saved as a file. Then the program must be compiled before it can be executed. To accomplish the same in LabVIEW, all we have to do is press the Run button.

SIMPLIFIED INTERACTIONS BETWEEN THE GRAPHICAL PROGRAM AND THE COMPUTER

It is appropriate at this point to take a brief look at how the program interacts with the computer operating system (OS), such as Microsoft Windows, and hardware components. Computers receive information from the external world and process, store, and communicate that information per instructions from the user. There are a wide variety of computing systems, ranging from supercomputers to personal digital assistants and embedded systems. This text focuses on the personal desktop unit (PC).

Figure 2.25 shows some typical hardware components of PC-based DAQ and their basic interaction. For example, memory can be cache, RAM (random access

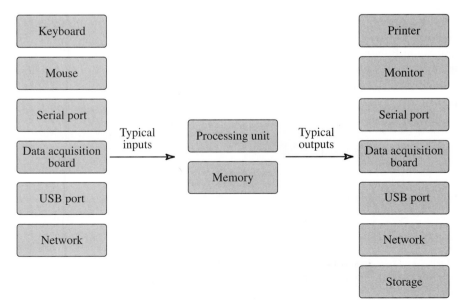

FIGURE 2.25
Example of computer component interactions

memory), or virtual memory (on the hard drive). Storage can be a hard drive, network drive, floppy disk, compact disk, zip disk, USB (universal serial bus) drive, and so on. Data acquisition boards can be inserted into the computer bus slots or attached to the serial or USB ports. For a complete discussion of interactions between, and organization of, computer components and their interactions with software, please refer to a computer organization text such as the one written by D. A. Patterson and J. L. Hennessy (2003).

Computers store and operate on binary values only and receive instructions from the OS in native instructions that the computer's CPU can directly execute. For a C program to run, it has to be linked and compiled. So, after the program is written, it must be compiled to produce an executable file. Some C development environments allow the user to test the program without compiling.

LabVIEW compiles the program at run time (when you click the Run Button). Consequently, if you want to run a LabVIEW graphical program on another computer, the computer has to have the LabVIEW software, or you have to create an executable file and load the run-time engine (available free from www.ni.com). We will work with computers where LabVIEW is loaded in this text, but it is not difficult to create executable files.

After the LabVIEW or C program is converted to an executable, it works with the OS to share the computer resources with other programs running on the computer. For example, LabVIEW requests that the OS allocate some memory for the Pi indicator and the Pi constant, as shown in Figure 2.26. Windows will find some memory for the Pi indicator, perhaps while it is reallocating memory for other operations. Then, the OS transfers the value of the Pi constant into the Pi indicator memory area when the program runs. Furthermore, LabVIEW requests that the OS display the Pi indicator and the initial and resulting values in the indicator on the computer monitor along with the front panel and block diagram windows.

Of course there are many other interactions with mouse moves, retrieving the value of π from a library, starting the processor, ending the program, and so on. The graphic in Figure 2.26 shows some obvious interactions specific to the Pi program that occur between the VI, the OS, and the computer memory. It is useful to understand that our programs must communicate with the OS to instruct the computer to perform the necessary actions.

FIGURE 2.26
Simplistic sequence of computer interactions

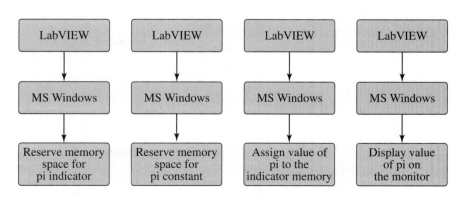

INTRODUCTION TO DATA TYPES

Data types are important because they provide information the operating system uses to determine the amount and type of memory to allocate. The LabVIEW Pi indicator representation (or data type) was defined when we placed the indicator on the front panel window. The indicator default representation was double-precision floating point.

Table 2.1 lists LabVIEW control and indicator data types, their block diagram icon, proper name, usage, and default value. If you are working in LabVIEW while reading this text, display this table and observe the colors assigned to each data type.

SIMPLE CALCULATIONS

Our next example program is a simple calculation of the area of a circle. Begin by opening a blank VI and building the front panel window shown in Figure 2.27, or begin with the Pi program and modify it.

Our program will allow the user to input a radius, and the program will calculate and display the area. In the Pi example program, we worked with an indicator because the user did not provide any input. Users input information to LabVIEW programs via controls. So we will need two objects on the user interface: a control and an indicator. The correct technique is to place controls on the left side of the user interface, and indicators on the right to follow English writing style. This style makes our program user friendly and reduces documentation.

Right-click on the Front Panel and choose the Numeric Controls sub-palette, as shown in Figure 2.27. Next, choose a Numeric Control and place it on the left side of the front panel window. Label it using the procedure previously described in the Pi example.

Use the technique from the Pi example to place an indicator on the right side of front panel window and label it as shown in Figure 2.27.

Graphical user interfaces (GUIs) look better if the objects are aligned. We can align the bottom edges of the front panel objects as shown in Figure 2.28 by selecting both and using the Align button from the toolbar. To select an object, move the cursor carefully and precisely onto the edge of the object until the cursor shape becomes an arrow, then click. A marquee will surround the object. Use a border of the control or indicator rather than the border of the label, since we can select the label separately to move it relative to the control or indicator. Hold the shift key down while selecting the second object. Alternately, we can draw a box around both objects with the cursor in the arrow shape (the select tool). After both are highlighted, click the Align button and choose Align Bottom Edges. Note the other options for aligning horizontally and vertically. Remember to save the program often while coding (using Ctrl-S), not just at the end. Save as circle area.vi.

All VIs should be user friendly, meaning they should be easy to use and intuitive. When the user views the program, the program functionality and use should be obvious. Alignment and placement enhance usability, but labels with carefully chosen words, size, and placement are also essential. There are two types of labels in Figure 2.28: owned and free. Owned labels are part of an object. The indicator label "Area

TABLE 2.1 LabVIEW Block Diagram Object Data Types (LabVIEW Help > Block Diagram Objects)

Control Indicator	Data Type	Use	Default Values
SGL	Single-precision, floating-point numeric	Saves memory and does not overflow the range of the numbers.	0.0
DBL	Double-precision, floating-point numeric	Is the default format for numeric objects.	0.0
EXT	Extended-precision, floating-point numeric	Performs differently depending on the platform. Use only when necessary.	0.0
CSG	Complex single-precision, floating-point numeric	Same as single-precision, floating-point, with a real and an imaginary part.	0.0 + 10.0
CDB	Complex double-precision, floating-point numeric	Same as double-precision, floating-point, with a real and an imaginary part.	0.0 + 10.0
CXT	Complex extended-precision, floating-point numeric	Same as extended-precision, floating-point, with a real and an imaginary part.	0.0 + 10.0
EXP	Fixed-point numeric	Stores values that fall within a user-defined range. This data type is useful when you do not need the dynamic functionality of floating-point representation or when you want to work with a target that does not support floating-point arithmetic, such as an FPGA target.	0.0
I8	8-bit signed integer numeric	Represents whole numbers and can be positive or negative.	0
I16	16-bit signed integer numeric	Same as above.	0
I32	32-bit signed integer numeric	Same as above.	0
I64	64-bit signed integer numeric	Same as above.	0
U8	8-bit unsigned integer numeric	Represents only non-negative integers and has a larger range of positive numbers than signed integers because the number of bits is the same for both representations.	0
U16	16-bit unsigned integer numeric	Same as above.	0
U32	32-bit unsigned integer numeric	Same as above.	0
U64	64-bit unsigned integer numeric	Same as above.	0
	<64.64>-bit time stamp	Stores absolute time with high precision.	12:00:00.000 AM 1/1/1904 (Universal Time)
	Enumerated type	Gives users a list of items from which to select.	—
TF	Boolean	Stores Boolean (TRUE/FALSE) values.	FALSE
abc	String	Provides a platform-independent format for information and data, which you can use to create simple text messages, pass and store numeric data, and so on.	empty string
	Array	Encloses the data type of its elements in square brackets and takes the color of that data type. As you add dimensions to the array, the brackets become thicker.	—
CDB	A matrix of complex elements.	The wire pattern differs from that of an array of the same data type.	—
DBL	A matrix of real elements.	The wire pattern differs from that of an array of the same data type.	—
	Cluster	Encloses several data types. Cluster data types appear brown if all elements in the cluster are numeric or pink if all elements of the cluster are of different types. Error code clusters appear dark yellow, while LabVIEW class clusters are crimson by default.	—
	Path	Stores the location of a file or directory using the standard syntax for the platform you are using.	empty path
	Dynamic	(Express VIs) Includes data associated with a signal and the attributes that provide information about the signal, such as the name of the signal or the date and time the data was acquired.	—
	Waveform	Carries the data, start time, and Δt of a waveform.	—
	Digital waveform	Carries start time, Δx, the digital data, and any attributes of a digital waveform.	—
	Digital	Encloses data associated with digital signals.	—
	Reference number (refnum)	Acts as a unique identifier for an object, such as a file, device, or network connection.	—
	Variant	Includes the control or indicator name, the data type information, and the data itself.	—
	I/O name	Passes resources you configure to I/O VIs to communicate with an instrument or a measurement device.	—
	Picture	Includes a set of drawing instructions for displaying pictures that can contain lines, circles, text, and other types of graphic shapes.	—

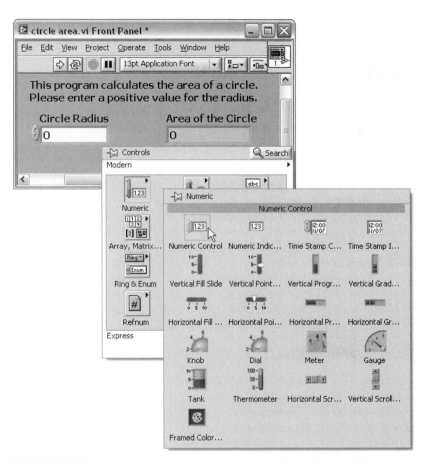

FIGURE 2.27 Building the front panel window for the circle area VI

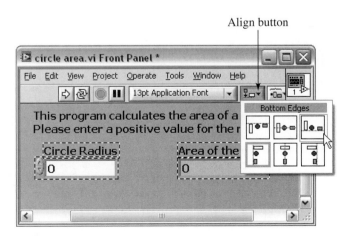

FIGURE 2.28 Aligning front panel objects for the circle area VI

of the Circle" is an example of an owned label. The label, "This program calculates the area of a circle. Please enter a positive value for the radius." is a free label.

Owned labels belong to and move with a particular object and annotate that object only. We can move an owned label independently by selecting it and excluding the object. We can hide owned labels but cannot copy or delete them independently of their owners. We can also display a unit label for numeric controls and indicators by selecting Visible Items >> Unit Label from the shortcut menu. Owned labels are similar to variable names in a text-based programming.

Free labels are not attached to any object. Free labels are useful for documenting code on the block diagram window and for listing user instructions on the front panel window. Double-click an open space or use the Labeling tool to create free labels or to edit either type of label.

The circle area program front panel label font is 18 size and bold. Use the Text Settings pull-down menu, shown in Figure 2.29, to change the attributes of text on either the front panel or block diagram windows. If we select objects or text before making a selection from the Text Settings pull-down menu, the changes apply to everything selected. If we select nothing, the changes apply to the default font. Changing the default font does not change the font of existing labels. It affects only those labels you create from that point forward. Select Font Dialog from the Text Settings pull-down menu on the Front Panel to apply specific font styles to selected text.

The Text Settings pull-down menu contains the following built-in fonts:

> Application Font—default font used for Controls and Functions Palettes and text in new controls
> System Font—used for menus
> Dialog Font—used for text in dialog boxes

FIGURE 2.29
Text attributes
pull-down menu

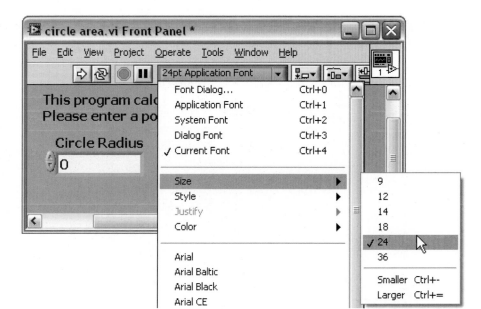

This text presents a very brief introduction to the complex GUI design issue (which is so complex it has been the subject of several texts). Some introductory suggestions for GUI design include the following:

> ➤ Give objects meaningful names. Use consistent capitalization.
> ➤ Place names (labels) consistently, for example, all left-justified.
> ➤ Use standard, consistent fonts throughout all front panel windows.
> ➤ Put default values in parentheses on labels.
> ➤ Include unit information in names; for example, use Time Limit (s).
> ➤ Arrange objects logically. Put the most important objects in the most prominent positions.
> ➤ Put inputs (controls) on the left and outputs (indicators) on the right.
> ➤ Arrange controls attractively, using the Align Objects and the Distribute Objects pull-down menus.
> ➤ Do not overlap controls.
> ➤ Use color logically and sparingly, if at all.
> ➤ Provide a Stop button if necessary. Do not use the Abort button to stop a VI. We will discuss this in more detail later when we cover repetition structures.

Further information is available in the LabVIEW Help >> Search the LabVIEW Help >> Contents Tab >> Development Guidelines, and in Ritter (2004).

ARITHMETIC IN LABVIEW

We will now expand the Pi programs concepts to calculate the area of a circle using two branches, one where the user provided a positive value of radius and another where the user mistakenly input a negative value.

We will begin by implementing and testing the positive-value branch, which requires inserting some arithmetic functions to compute the area $= \pi r^2$ calculation. Insert the Multiply Function from the Arithmetic subpalette (Functions >> Numeric), as shown in Figure 2.30. Place the multiply function close to the radius control to enable an automatic wire, and add a wire to compute radius squared. Make the branch in the wire by moving the cursor close to the wire (the wire will blink). Left-click with the wiring tool to start the new wire, extend it to the terminal, and left-click again to make the connection and end the wire.

Next, duplicate the Multiply Function and place it as shown in Figure 2.30. There are several ways to duplicate an object in LabVIEW:

1. Select the object and then use the edit menu or shortcut keys (Ctrl-C to copy and Ctrl-V to paste).
2. Hold the control key down while selecting the object, and drag it to the new location.
3. Leave the Functions Palette window open and get a duplicate object.

Insert the Pi constant as we did in the previous example program (unless you are modifying your existing Pi program). Finally, complete the wiring as shown in Figure 2.30, save the program, and test it. First, input a radius in the control,

FIGURE 2.30
Placing the Multiply
Function on the block
diagram window.

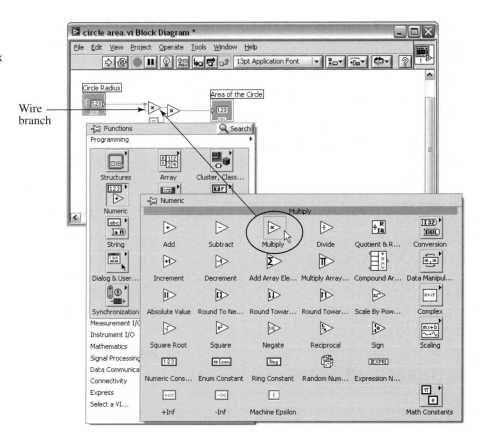

and then select the Run button. What should be displayed as the area if the user enters a value of 1? 0? −1? We haven't added the branch for negative values yet, so the program will report a meaningless area when the user enters a negative value.

If you are working in LabVIEW while reading this text, notice that the wires, front-panel object icons, and the Pi constant are all orange in color indicating consistent double-precision floating-point data type in the VI.

DATA FLOW

A big difference between LabVIEW and text-based programming (e.g., C) is that statements typed in text programs execute sequentially, top to bottom (with some exceptions caused by control structures), but placing LabVIEW icons above or below, right or left, doesn't specify the execution order. If you place icon A to the left of and/or above icon B, icon B might execute before icon A. Execution order in LabVIEW is controlled by a concept called data flow. In data flow, an icon or function does not execute until it has data into all of its inputs.

The following sequence explains the data flow in the circle area VI:

1. When the Run button is pressed, the value of data contained in the Circle Radius control flows to the first multiply operator.
2. The first multiply operator executes because its inputs have values, and the value computed by the first multiply operator flows to the second multiply operator.
3. The Pi constant value flows to the second multiply operator. (Operations 2 and 3 could be reversed. It is the computer's choice. But it doesn't matter. The important point is that the second multiply operator won't execute until it receives values from both upstream processes.)
4. The second multiply operator executes after receiving both input values. The value computed by the second multiply operator flows to the Area of the Circle indicator.

DEBUGGING

We can examine data flow further with the LabVIEW Execution Highlighting tool. Execution Highlighting animates the execution of the Block Diagram, showing the movement of data from one node to the next using bubbles that move along the wires. Activate Execution Highlighting by toggling the button with the light-bulb icon on the block diagram menu bar, and then run the program using the Run button on the block diagram menu bar, as shown in Figure 2.31. Execution Highlighting is also an excellent debugging tool.

Single Stepping is another debugging tool. Activate Single Stepping by toggling the Pause button on the menu bar, then use the Step Into, Step Over, and Step Out Of buttons. Single Stepping allows you to view each action of the VI as it runs. If you

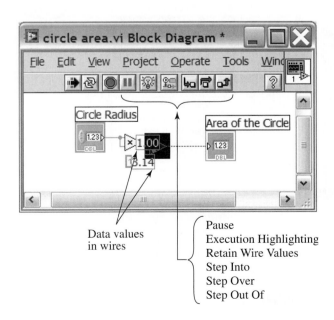

FIGURE 2.31 Execution Highlighting (program paused during execution of second multiply function)

Data values in wires

Pause
Execution Highlighting
Retain Wire Values
Step Into
Step Over
Step Out Of

are programming in LabVIEW while studing the text, you will notice that the Pause button vertical lines turn red when the VI is paused.

Step Into opens a node and pauses. When you click the Step Into button again, it executes the first action and pauses at the next action of the subVI or structure. Step Over executes a node and pauses at the next node. Step Out finishes executing the current node and pauses. Nodes blink to indicate they are ready to execute.

Probes are yet another excellent debugging tool. Right-click on a wire and choose Probe from the shortcut menu to display values in the wires while the program runs with or without Execution Highlighting, as shown in Figure 2.32. You can place several probes on a block diagram window. LabVIEW inserts a numbered probe glyph on the wire and displays the value in a numbered floating probe window to distinguish between probes. You can use the Probe tool in combination with Execution Highlighting, Single Stepping, and Breakpoints. If you want a probe to display the data that flowed through the wire at the last VI execution, use the Retain Wire Values option.

Figure 2.32 shows the use of the generic probe. You can create a custom probe by right-clicking a wire and selecting Custom Probe from the shortcut menu, as shown in Figure 2.33. A floating Probe window appears. If you have not yet selected a supplied probe or created a custom probe, LabVIEW finds a probe that matches the data type of the wire you right-clicked.

If you right-click a wire and select Custom Probe, you have several options. You can choose a probe based on one of the controls from the Controls Palette—a chart or graph, for example—or you can create a new probe. If you select New, a dialog box opens, giving you the option of creating an entirely new probe or of creating a probe based on an existing probe. LabVIEW stores the supplied probes in the labview\vi .lib_probes directory and stores custom probes in the directory you specify.

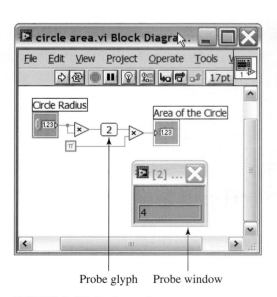

Probe glyph   Probe window

FIGURE 2.32 Probe tool

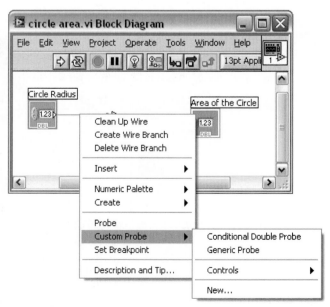

FIGURE 2.33 Create a custom probe

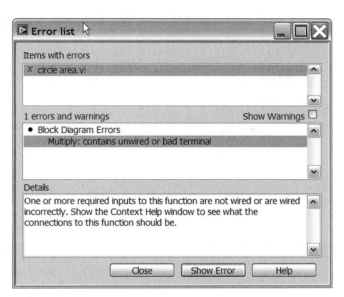

FIGURE 2.34
Broken Run button   **FIGURE 2.35** Error list from clicking a broken Run button

To use the probe you created, right-click the wire again and select Probe from the shortcut menu. After you create a probe, that probe becomes the default probe for that data type, and LabVIEW loads that probe when you right-click the wire and select Probe from the shortcut menu. To change the default probe for a data type, right-click a wire of that data type, select Custom Probe from the shortcut menu, and select a supplied, custom, or generic probe in the shortcut menu.

If the program runs but gives an incorrect answer, Execution Highlighting, Single Stepping, and Probes are excellent debugging tools. But, if the program won't run, you can't use them. However, the Run button becomes an excellent debugging tool in the instance where a program won't run. If a VI does not run, it is broken, or nonexecutable. The Run button appears broken, as shown in Figure 2.34.

Click the broken Run button and a window will pop up that describes the errors, as shown in Figure 2.35. Double-click an error in the list, or highlight it and select the Show Error button at the bottom of the window, and LabVIEW will highlight the location of the error on the block diagram window. The Help button at the bottom of the window will provide help on fixing the error.

ONLINE CONTEXT HELP

It is easy to get help while you are coding LabVIEW programs. Move the cursor over an object, such as the Multiply Function, and press Ctrl-H, or right-click to bring up the shortcut menu and choose Help. The Context Help window pops up and will change when the cursor is moved over another object. The Help window provides a lot of useful information, as shown in Figure 2.36. Choose the ? for help to obtain more detailed information. You can lock the Help window on an object by clicking

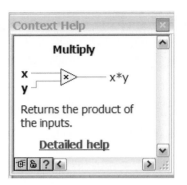

FIGURE 2.36 Context help for the Multiply Function

on the lock symbol in the lower-left corner of the window. In addition to describing the function or VI, the Help window contains information for wiring, such as the terminal type. There are three types of terminals: required, recommended, and optional. Bold font indicates a required terminal. Notice that both **x** and **y** input terminals must be wired on the Multiply Function.

UNITS

LabVIEW supports units on the block diagram and front panel. Some constants have units associated with their values, the speed of light (c) for example. If we place the c constant on the block diagram window and right-click the terminal to create an indicator, as shown in Figure 2.37, the Units label automatically appears on the indicator as s^-1 m.

Notice that EXT appears on the bottom of the indicator icon on the block diagram window indicating the default data type for the c constant is extended-precision floating-point.

LabVIEW supports a large number of abbreviations for units. Table 2.2 lists LabVIEW base units in the SI system. Table 2.3 shows SI derived units that LabVIEW recognizes. Table 2.4 lists the non-SI units recognized by LabVIEW.

FIGURE 2.37
Display an indicator for c that automatically displays units

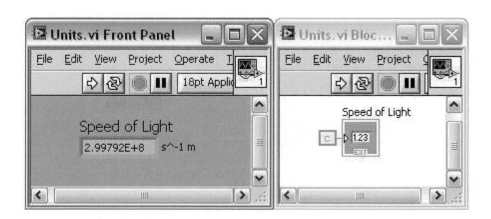

TABLE 2.2   LabVIEW Base Units (from LabVIEW Help)

Quantity Name	Unit	Abbreviation
Length	meter[*]	m
Mass	kilogram[*]	kg
Time	second[*]	s
Electric current	ampere[*]	A
Thermodynamic temperature	kelvin[*]	K
Amount of substance	mole[*]	mol
Luminous intensity	candela[*]	cd

TABLE 2.3   LabVIEW Units Derived from the SI Base Units (from LabVIEW Help)

Quantity Name	Unit	Abbreviation
Area	square meter	m^2
Volume	cubic meter	m^3
Mass	gram	g
Velocity	meters per second	m/s
Acceleration	meters per second squared	m/s^2
Wave number	reciprocal meter	m^{-1}
Mass density	kilograms per cubic meter	kg/m^3
Specific volume	cubic meters per kilogram	m^3/kg
Current density	amperes per square meter	A/m^2
Magnetic field strength	amperes per meter	A/m
Amount-of-substance concentration	moles per cubic meter	mol/m^3
Luminance	candelas per square meter	cd/m^2
Mass fraction	kilogram per kilogram	kg/kg
Plane angle	radian[*]	rad
Solid angle	steradian[*]	sr
Frequency	hertz	Hz
Force	newton	N
Pressure, stress	pascal	Pa
Energy, quantity of heat, work	joule	J
Power, radiant flux	watt	W
Electric charge	coulomb	C

(Continued)

TABLE 2.3   *Continued*

Quantity Name	Unit	Abbreviation
Electric potential difference	volt	V
Capacitance	farad	F
Electric resistance	ohm	Ohm
Electric conductance	siemens	S
Magnetic flux	weber	Wb
Magnetic flux density	tesla	T
Inductance	henry	H
Temperature (absolute)	degree Celsius	degC
Luminous flux	lumen	lm
Illuminance	lux	lx
Activity (of a radionuclide)	becquerel	Bq
Absorbed dose, kerma, specific energy (imparted)	gray	Gy
Dose equivalent	sievert	Sv

TABLE 2.4   **LabVIEW Non-SI Units (from LabVIEW Help)**

Quantity Name	Unit	Abbreviation
activity	curie	Ci
area	acre	acre
	barn	b
	hectare	ha
dynamic viscosity	poise	P
electric power	horsepower (electric)	hp
energy	British thermal unit (mean)	Btu
	calorie (thermal)	cal
	electron volt	eV
	erg	erg
force	dyne	dyn
	pound-force	lbf
illuminance	foot-candle	fc

TABLE 2.4 *Continued*

Quantity Name	Unit	Abbreviation
length	ångström	Angstrom
	astronomical unit	AU
	chain	chain
	cubit	cubit
	fathom	fathom
	fermi	fermi
	foot	ft
	furlong	furlong
	hubble	hubble
	inch	in
	light year	ly
	mile	mi
	rod	rod
	yard	yd
lens power	dioptre	dpt
magnetism	emu	emu
	gauss	Gs
	maxwell	Mx
	oersted	Oe
mass	carat	CM
	dram	dr
	grain	gr
	metric tonne	t
	ounce	oz
	pound-mass	lb
	slug	slug
	stone	stone
	ton (UK)	Ton
	ton (US short)	ton
	unified atomic mass unit	u
plane angle	degree	deg
	minute	'
	second	"
pressure	atmosphere	atm
	bar	bar
	meter of mercury	m.Hg
	pound per square inch	psi
	torr	torr

(Continued)

TABLE 2.4   *Continued*

Quantity Name	Unit	Abbreviation
radiation dose	Rad	Rad
radiation exposure	roentgen	r
radiation exposure: man	roentgen equivalent man	rem
temperature (absolute)	degree Fahrenheit	degF
temperature difference (relative)	Celsius degree	Cdeg
	Fahrenheit degree	Fdeg
time	day	d
	hour	h
	minute	min
	year	a
velocity	knot	kn
volume	barrel	barrel
	bushel	bushel
	fluid dram	fl.dr
	fluid ounce	fl.oz
	gallon (imperial)	UKgal
	gallon (US)	gal
	gill	gill
	liter	l
	minim	minim
	peck	peck
	pint	pint
	quart	quart

SUMMARY

➤ To start LabVIEW, select start >> Programs >> National Instruments >> LabVIEW 8.5

➤ The Getting Started window allows the user to select from a menu bar and four buttons on the right, for a new VI, help, open an existing VI, or other tasks.

➤ When we open a new VI, two windows appear: the front panel window, which is the graphical user interface (GUI), and the block diagram window, which is where we will develop the code.

➤ We build a GUI by using a set of tools to insert and manipulate objects on the front panel window. LabVIEW uses two fundamental objects called controls and indicators to input and output information.

➤ To place an object on the front panel window, move the cursor onto the front panel window and right-click to display the Controls palette.

➤ Pinning the palette adds a three-button menu, including a Search button. Search is very useful for locating objects quickly in LabVIEW's large tree of functions.

➤ When we place an object on the front panel window, its title, or label, has a black background.

Before we make any additional moves or clicks with the mouse, we can change the title of the control easily by typing a new name.

➢ The LabVIEW cursor automatically switches between different shapes (tools) to perform different operations.

➢ Shortcut keys, such as Ctrl-S for frequent saving, facilitate programming in LabVIEW.

➢ When we place an object on the front panel window, it automatically appears as an icon on the block diagram window.

➢ The palette used for block diagram objects is called the Functions palette.

➢ We connect objects on the block diagram window by extending wires between them with the wiring tool.

➢ To run a VI, click the Run button, which is the white arrow on the left side of the front panel toolbar. We do not have to link and compile before running.

➢ LabVIEW graphical programming doesn't require header files, preprocessor directives, rules for variable names, variable definitions, format instructions, and special punctuation such as semicolons and braces. LabVIEW automatically includes these details, and it also gives the developer capabilities to change the automatic defaults when necessary.

➢ LabVIEW automatically defines data types when we place objects on the front panel or block diagram windows, and it gives us the capability to change the default types if necessary.

EXERCISES

1. In addition to making notes of the material in this chapter, add a list of shortcut keys that were introduced. Put it on a separate page in your notebook and update it as information is added in subsequent chapters.

2. Read the text carefully and build the Pi program example while you are reading and then test it.

3. Build the circle area example program either by beginning a new program or by modifying the Pi program. Negative radius input values are okay at this point. We will explain how to eliminate them later.

4. Build the VI in Chapter 1 of the Getting Started with LabVIEW pdf manual. It can be found in either of two locations. One location is in the LabVIEW Help. Help >> Search LabVIEW Help >> Contents Tab >> Technical Support and Professional Services >> LabVIEW Documentation Resources. Under Print Documents, select the Getting Started with LabVIEW manual. The other location is in the Windows Start Programs Menu: Start >> All Programs >> National Instruments >> LabVIEW 8.5 >> LabVIEW Manuals >> LV_Getting_Started.pdf

5. Demonstrate the following:
   a. Add the alignment grid to the block diagram window (use Help to find the alignment grid).
   b. Align the horizontal center of two objects on the block diagram window.

   c. Search the Functions palette for the Gravitational Constant.
   d. Show the Context Help window for the Gravitational Constant.

6. Create a graphical program, similar to the Pi program, but have it display many more constants. Align the objects on the front panel window so the GUI is very user friendly. Use appropriate labels on the indicators for the proper name and units of each constant. Display as many constants as you can find in the Functions palette. Expand the size of the indicators so the values are completely shown. Some constants may require that you use the unit label as well as the name label on the indicator. The labels will appear automatically if you right-click on the constant output terminal and choose create >> indicator. Refer to the information on Numeric Units and Strict Type Checking in the LabVIEW Help.

7. Predict the execution order by predicting which constant's value will be displayed first and which will be displayed last for the previous question.
   First Displayed: _____   Last Displayed: _____
   Then run the program with Execution Highlighting activated, and compare the actual execution order with your prediction.
   First Displayed: _____   Last Displayed: _____

8. Experiment with debugging.

a. Open the circle area VI and resave it as circle area mistake VI. Remove one of the input wires to the multiply function. What happens to the Run button?

_____

b. Click the Run button even though it is broken. What is displayed?

_____

c. Assume you were programming quickly and used the Addition Function instead of the Multiply Function. Right-click the Multiply Function and choose Replace >> Numeric Palette >> Addition from the shortcut menu. The program runs but gives incorrect answers when you test it with a radius of 1. Use a probe to display the values in the wires. Use Execution Highlighting to display the values in the wires.

9. Build a new program that displays c in m/s, ft/min, light years/year, and mph in four different indicators. Right-click on the units label and choose Build Unit String to change the units for each indicator. Select Help in the lower-right corner to learn how to delete the old units and add new ones.

10. Create a graphical program that calculates and displays the cross-sectional area inside a tube in mm^2, the cross-sectional area of the tube wall in mm^2, and the tube outer diameter in mm. The user will input the inner diameter and the tube wall thickness in mm.

Data Acquisition

INTRODUCTION

Chapter 2 introduced the software development portion of the data acquisition system. This chapter introduces the hardware, including interfacing to plug-in boards, signal conditioning, use of the BNC-2120 signal accessory, driver and configuration software, and develops an example program that acquires temperature data. We will begin to develop more sophisticated applications in Chapter 4 by introducing software design, selection, repetition, and timing.

OUTLINE

THE TEMPERATURE MEASUREMENT PROGRAM

Now that we have learned how to develop and run a simple program, we can learn how to automatically measure physical properties. Making measurements automatically with a computer is called data acquisition (DAQ).

In this section, we will create the Temperature Measurement Program, shown in Figure 3.1.

It can be created quickly with the following steps:

> Open LabVIEW.
> Open MAX.
> Configure devices in MAX.
> Create the DAQ Assistant on the block diagram window.
> Configure the measurement.
> Create the Multiply Function and scaling constant on the block diagram window.
> Create the chart indicator on the front panel window.
> Wire the DAQ Assistant, Multiply Function, scaling constant, and chart indicator.
> Save, run, test the program.

We will discuss each of the steps in detail in the following.

We will use a temperature transducer located in the BNC-2120 accessory that is plugged into the PCI-6024E data acquisition card that resides in our computers, as shown in Figure 3.2. The instructions that follow use the PCI-6024E as an example, but other NI-DAQ devices and simulated devices can be used as well.

FIGURE 3.1
Temperature
Measurement
Program

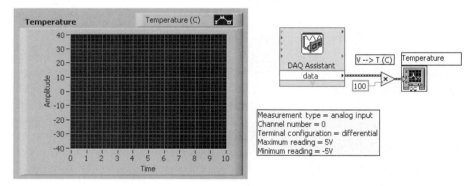

FIGURE 3.2
Typical data
acquisition system

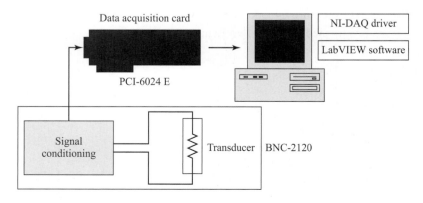

BNC-2120 DAQ ACCESSORY

We chose the BNC-2120 for this text because it has a lot of functionality for a low cost. The text is written so that other devices can be used in its place, including simulated devices. The BNC-2120 DAQ accessory, shown in Figure 3.3 and Figure 3.4, is named after the BNC connectors used for analog input connections. BNC means Bayonet Nut Connector, which resembles the way a bayonet connects to the end of a rifle. It is a commonly used plug and socket connection for instrumentation, audio, and networking with coaxial cables.

For the simple DAQ application in this chapter, the BNC-2120 connects to a PCI-6024 E plug-in DAQ board with a 68-pin input/output (I/O) connector and cable. The manual for the BNC-2120 is available online: www.ni.com>Support> Product Reference>Manuals>BNC-2120.

FIGURE 3.3 Photograph of the National Instruments BNC-2120 accessory

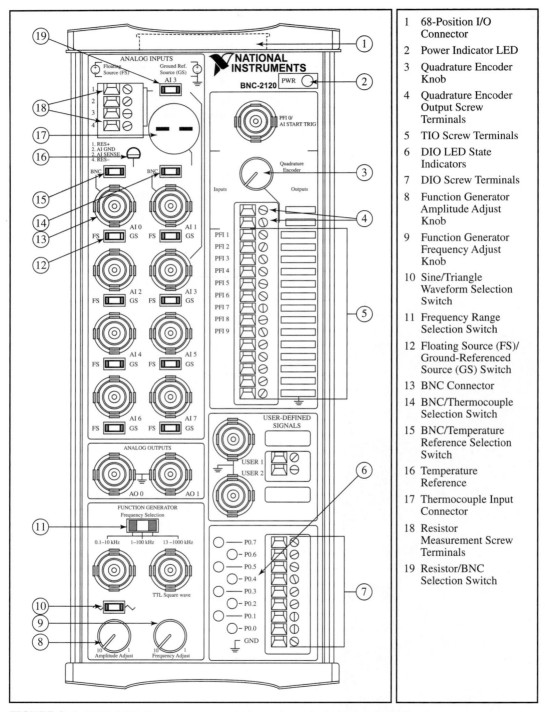

FIGURE 3.4 BNC-2120 Components, controls, and terminals (www.ni.com)

MEASUREMENT & AUTOMATION EXPLORER

Before we write a program to measure temperature, we need to provide information to the computer operating system so it can establish communication with a transducer, in this case with the temperature transducer in the BNC-2120 through the PCI-6024E DAQ board. We don't need to do this each time we write a program, just when using the devices for the first time or when we change devices.

We provide information to the operating system with the Measurement & Automation Explorer (MAX) Program. We will use MAX to view devices and instruments connected to a computer and configure hardware and software. But, MAX can be used for other tasks, too.

Open MAX from the Start Programs Menu (Start >> Programs >> National Instruments >> Measurement & Automation Explorer) or from LabVIEW (Tools >> Measurement & Automation Explorer).

Expand the Devices and Interfaces as shown in Figure 3.5. If you expand the NI-DAQmx Devices and see the PCI-6024E, the Windows PC has recognized the DAQ board. If your device does not appear, press <F5> to refresh the view in MAX. The PC will assign a device number, usually "Dev 1," to the board. It is possible to have several devices, or several plug-in boards, that will have different device numbers for complex measurement and control systems.

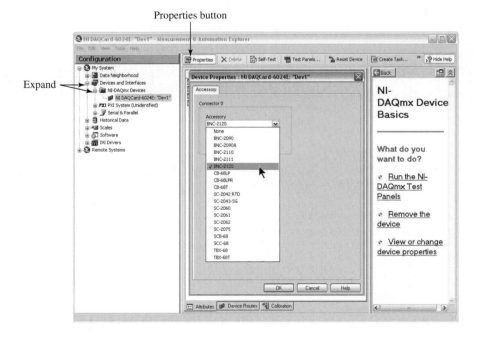

FIGURE 3.5
Measurement &
Automation Explorer

Next click the Properties button. Choose BNC-2120 in the Accessory pull-down, as shown in Figure 3.5. Exit MAX from the File menu, and the configuration changes will be saved.

If your computer doesn't have DAQ hardware such as a plug-in board, you can simulate devices in MAX by right-clicking Devices and Interfaces and choosing Create New . . . , as shown in Figure 3.6.

This opens the window shown in Figure 3.7. Choose the NI-DAQmx Simulated Device.

MAX opens the window shown in Figure 3.8. Choose one of the devices, such as NI PCI-6024E, as shown.

FIGURE 3.6 Create a simulated device in MAX

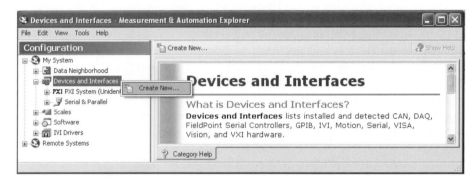

FIGURE 3.7 Create new device window in MAX

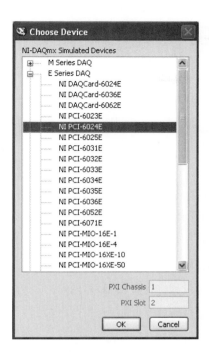

FIGURE 3.8 Choose device window in MAX

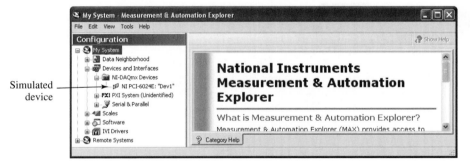

FIGURE 3.9
Simulated device configuration in MAX

Simulated device

The resulting configuration in MAX is shown in Figure 3.9. If you are working in LabVIEW while reading this text, you will notice that the simulated device icon in MAX is yellow to distinguish it from actual hardware whose icon is white.

DAQ ASSISTANT

After configuring with MAX, the computer can communicate with the PCI-6024E and the BNC-2120, so we can begin developing our program. Open a blank VI in LabVIEW, and expose the block diagram window. Add the DAQ Assistant (Functions >> Measurement I/O >> NI-DAQmx >> DAQ Assistant) to the Block Diagram, as shown in Figure 3.10.

FIGURE 3.10
DAQ Assistant

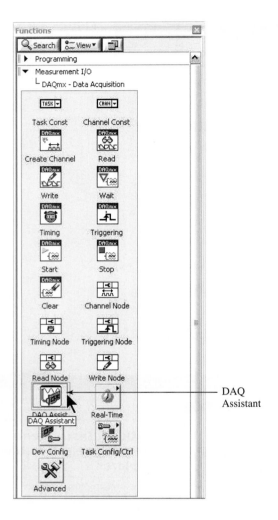

The DAQ Assistant is a graphical interface (called an express VI) for configuring measurement tasks, channels, and scales. The DAQ Assistant opens the window shown in Figure 3.11 giving us the option of choosing a measurement type. Our measurement is an input signal, so expand the Acquire Signals item. It is an analog value because it is a continuously varying electrical signal that represents temperature, so expand Analog Input.

The other signal types are counter, digital, and TEDS. Counters accumulate pulses, and digital measurements are for signals that are either 1 or 0. Neither is appropriate for our temperature measurement. They are appropriate for pulses such as those produced by optical encoders or rotary flowmeters per the information in Table 1.1. TEDS (transducer electronic data sheet) facilitates sensor plug and play installation. With TEDS, the sensor identifies and describes itself to the data acquisition system to which it is connected, using data stored on the sensor.

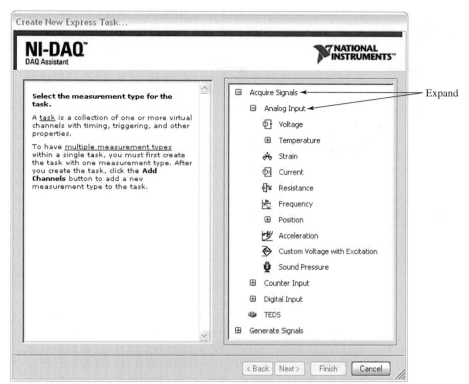

FIGURE 3.11
Measurement type selection in the DAQ Assistant

After expanding Analog Input, we can select Temperature or Voltage. Temperature would help us if we had a difficult conversion from our electrical signal units to temperature units, for example, if we were using thermocouples or thermistors. However, the BNC-2120 outputs the temperature-transistor signal in volts, and it is easy to convert to Celsius, so we will choose Voltage. After choosing Voltage, the DAQ Assistant window shown in Figure 3.12 opens. The PCI-6024E (device 1) has connections for as many as 16 sensor signals. Each signal is assigned to a channel. A LabVIEW program can receive data from all 16 channels, but the program is simpler and more efficient if we acquire the data from only the channel that is connected to this single temperature transducer. The BNC-2120 temperature transducer (Temperature Reference, or item 16, in Figure 3.4) is connected to AI (analog input), channel 0.

The DAQ Assistant window in Figure 3.12 uses the terms *virtual channels* and *physical channels*. A physical channel is a hardware terminal or pin at which a signal is connected. A virtual channel is a collection of property settings that can include a name, a physical channel, input terminal connections, the type of measurement or generation, and scaling information. Virtual channels are integral to every NI-DAQmx measurement.

Channels 0 and 1, AI 0 and AI 1, on the BNC-2120 have dual functionality. The BNC has switches for selecting either the BNC connectors, used for connecting external signals, or the internal temperature reference and the thermocouple connector.

FIGURE 3.12
Select the analog
input channel for the
DAQ Assistant

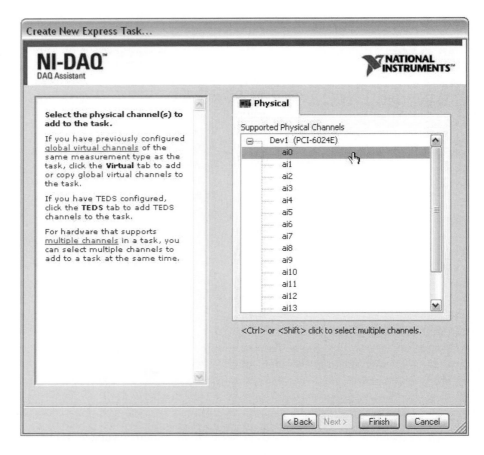

These switches (items 14 and 15 in Figure 3.4) are located above the BNC connectors for AI 0 and AI 1 (item 13 in Figure 3.4). Set the switch above AI 0 for the internal temperature transducer. The signal voltage of this transducer is linearly proportional to temperature: $T = V$ transducer $* 100$ where T is in degrees Celsius and V is in Volts. It is accurate to $\pm 1.5°C$. The switch above AI 1 supports a thermocouple input.

After specifying an analog input voltage signal on channel 0, we can specify how to make the measurement by defining the task with the window shown in Figure 3.13.

An NI-DAQmx task is a collection of one or more virtual channels with timing, triggering, and other properties. A task represents a measurement or a generation. Virtual channels can be part of a task or separate from a task.

There is a Run button and graph at the top of the Task window to test the task before closing the DAQ Assistant Task Configuration window. There is a Help window on the right side of the task configuration window.

The channel list box, on the left of the Task window, shows the voltage input channel and gives us the opportunity to delete, edit, or modify it or add additional channels. To the right of this box, we can set the Input Range in the Settings tab box. The default is -5 to $+5$ V. The measurement range setting is important because if we

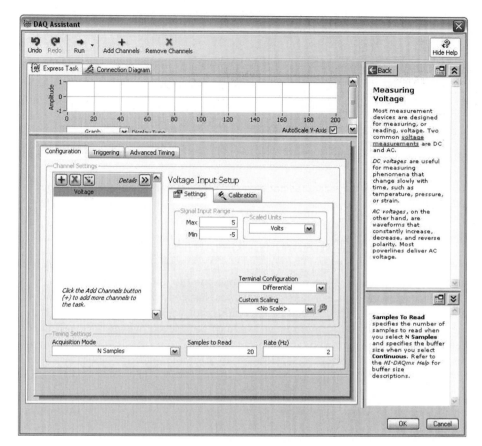

FIGURE 3.13
Defining the analog
input task in the DAQ
Assistant

set it too high, we increase error in the measurement, and if we set it too low, our measurement is out of the range. The error is caused by digitizing resolution, explained in the next section.

DIGITIZING RESOLUTION

The PCI-6024E converts the analog voltage signal to digital values for the computer. The DAQ board must convert the analog values in the input stream, like 5 V, to binary 0's and 1's (101, for example). This introduces an error because there aren't enough binary values to represent all possible analog values.

Figure 3.14 graphically depicts a digitizing resolution error emphasized by using a 3-bit resolution over a 10 V range. The DAQ board captures a voltage value at a point in time and converts that value to a binary number. This process changes the continuous input signal into a series of discrete points that are stored in the computer. To emphasize the point, assume we could purchase a DAQ board with a 3-bit resolution. Three bits have values ranging from 000 to 111, or 0 to 7. So the 3-bit ADC (analog to digital converter) in the DAQ board would have to convert the

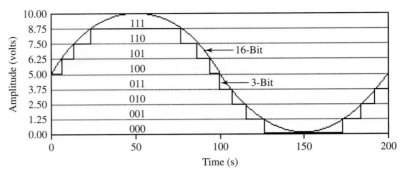

FIGURE 3.14 Digitizing resolution graph for 3-bit and 16-bit ADCs (from LabVIEW Help)

voltage signal to one of only eight possible values. The actual voltage is matched with the closest binary value, or in other terms, put into the closest bin.

This is similar to measuring with a ruler that has eight divisions per inch, and the object you measure falls between two divisions, but you had to report the measurement in 1/8-inch increments. An error would be introduced, which is the difference between the actual length and the division you choose to report. If you have a higher resolution ruler with marks every 1/16 inch, for example, the error would be less.

If we attempted to digitize a continuous analog 0–10 V sinusoid signal with our 3-bit ADC, we would represent 0 V as 000 and 10 V as 111, so we would divide the range, 10 V, by the number of possible values, 8, to calculate 1.25 V per division, as shown in Figure 3.14.

The ADC translates each measurement into one of its eight steps by selecting the step that is closest to the actual value of voltage. So, an actual measurement of 2 V would be converted to the closest value, 010, which is equivalent to 2.5 V, a 0.5 V resolution error! Consequently, we store a step function, shown in Figure 3.14, in the computer instead of the continuous function.

Higher-resolution ADCs, like the 16-bit shown, better approximate the actual measurement, but they still introduce a small error. Higher-resolution boards are more expensive. For example, the 18-bit resolution PCI-6024E costs $700. But reasonable resolution is available at lower prices with the student kits. The $159 USB-6008 Student Kit offers 10 kS/s at 12-bit resolution, and the $269 USB-6009 Student Kit delivers 48 kS/s at a resolution of 14 bits. Price is not completely determined by resolution and sample rate. Other factors such as number and type of input and output channels are also important considerations.

The PCI-6024E board in our computers has a digitizing resolution of 12 bits. Therefore, if the input range is set at ± 10 V, the digitizing resolution is $20\text{ V}/2^{12} =$ 4.88 mV. We can calculate the digitizing resolution for other input ranges with the following equation:

$$(\text{Input Voltage Range})/(2^{12}) = \text{Digitizing Resolution}$$

where 2^{12} = the number of discrete binary values available for each measurement

TABLE 3.1 **Software-Selectable Input Ranges and Resolution for the NI PCI-6024E DAQ board (modified from the PCI-6024E specification on ni.com.)**

Input Range	Resolution
−10 to +10 V	4.88 mV
−5 to +5 V	2.44 mV
−500 to +500 mV	244.14 μV
−50 to +50 mV	24.41 μV

The input voltage range is selected from Table 3.1, which shows the ranges allowed with the PCI-6024E DAQ board. Choose the smallest range that encompasses the maximum signal output.

To determine the range for our measurement, recall that we convert the voltage signal from the BNC-2120 temperature transducer with the sensitivity factor, 100°C /V. So, the 4.88 mV resolution converts to a resolution in temperature measurement of 0.488°C:

$$T(°C) = 100°C/V * 0.00488 \text{ V} = 0.488°C.$$

We might be able to improve this resolution of approximately ½ degree, because the −10 to +10 V range, is in the denominator of the digitizing resolution equation. Because we are measuring the temperature of ambient air, our measurements will probably be within −40 to +40°C, and in our computer lab, they will certainly be less than that! We can rearrange the above units-conversion equation to calculate the corresponding voltage range of the signal.

$$\Delta V = \Delta T/(100°C/V) = 80°C/(100°C/V) = 0.8 \text{ V}$$

where ΔV is the voltage signal range and ΔT is the temperature measurement range. So, referring to Table 3.1, we can certainly use a smaller range, say, −5 V to +5 V for a more general measurement application and −500 to +500 mV for a laboratory-specific measurement, which will reduce the resolution to 0.244°C or 0.024°C, respectively.

TERMINAL CONFIGURATION

Next, we will set the terminal configuration, which specifies how the measured signal relates to electrical ground. A grounded signal source is one in which the voltage signals are referenced to a system ground, as shown in Figure 3.15. Because such sources use the system ground, they share a common ground with the measurement device. The most common examples of grounded sources are devices that plug into a building ground through wall outlets, such as signal generators and power supplies. The grounds of two independently grounded signal sources generally are not at the same potential. The difference in ground potential between two instruments connected to the same building ground system is typically 10 mV to 200 mV. The difference can

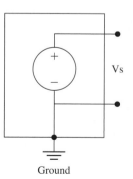

FIGURE 3.15
Grounded measurement (the LabVIEW Measurements Manual at ni.com)

be higher if power distribution circuits are not properly connected. This causes a phenomenon known as a ground loop. Refer to the LabVIEW Measurements Manual on ni.com for more information on this topic.

Referenced (RSE) and nonreferenced single-ended measurement (NRSE) systems are similar to grounded sources. A referenced single-ended measurement system measures voltage with respect to the ground, AIGND, which is directly connected to the measurement system ground. In a NRSE measurement system, all measurements are still made with respect to a single-node analog input sense (AISENSE on E Series devices), but the potential at this node can vary with respect to the measurement system ground (AIGND). A single-channel NRSE measurement system is the same as a single-channel differential measurement system.

We can make measurements on signals that are not referenced to ground, as shown in Figure 3.16. These measurements are called differential or floating. Since multiple channels can't share a common AIGND terminal, differential measurements require two channels, reducing the capacity of the PCI-6024E from 16 to 8 input signals.

Figure 3.17 compares differential and referenced single-ended measurements on a bridge circuit.

The default terminal configuration for the BNC-2120 temperature measurement is differential. Use the switch on the BNC-2120 (item 12 in Figure 3.4) to select either the floating or ground-referenced measurements. To avoid ground loops, use the ground-referenced source position labeled GS.

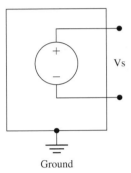

Ground

FIGURE 3.16
Floating measurement (after the LabVIEW Measurements Manual at ni.com)

Finish the Temperature Measurement VI

Next set the sampling information in the DAQ Assistant to N samples, 20 samples to read at a rate of 2 Hz, as shown in Figure 3.13, and press the OK button. Then save the program as Temperature Measurement.vi.

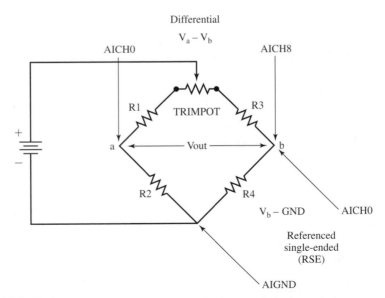

FIGURE 3.17 Comparison of differential and referenced single-ended measurements

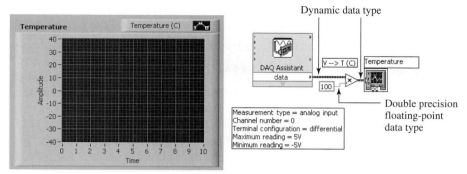

FIGURE 3.18
Temperature
measurement VI

Dynamic data type

Double precision
floating-point
data type

Plot legend window

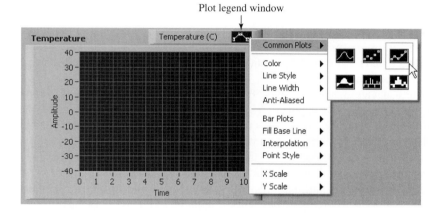

FIGURE 3.19
Changing the Plot
Style

Now we are ready to complete our temperature measurement program. Wire the data out from the DAQ Assistant to a Multiply function, as shown in Figure 3.18. The BNC-2120 temperature transducer output signal is a voltage. It is easily converted to temperature units (°C) by scaling by 100. Wire a numeric constant of 100 to display the result on the front panel, right-click on the output terminal of the Multiply function, and choose create >> graph indicator. Double-click on the block diagram and type the free labels shown. Save the program, run, and test it.

We can plot individual data points with the plot line on the Chart by changing the Plot Style, as shown in Figure 3.19. Right-click on the Plot Legend window, select Common Plots, and choose the style that displays both points and a line.

DYNAMIC DATA TYPE

As shown in Figure 3.18, the output signal from the DAQ Assistant is the dynamic data type. The dynamic data type is for use with Express VIs. Most other VIs and functions that ship with LabVIEW do not accept this data type. In addition to the data associated with a signal, the dynamic data type includes attributes that provide information about the signal, such as the name of the signal or the date and time the data was acquired. Attributes specify how the signal appears on a graph or chart.

SELECTING A DAQ DEVICE FOR ANALOG INPUT

We can generalize the information from the temperature measurement example and use it to guide hardware selection. If we want to purchase a DAQ device or evaluate a device for a new application, we should consider the number of channels, sampling rate, resolution, and input range. The initial step is to determine the physical properties we need to measure now and in the future. Then we select transducers to make the measurements. After transducers are selected, we can choose DAQ devices to acquire the signals. It is possible to select devices that are multifunctional; that is, they have channels for analog input, analog output, digital I/O, and counter/timers. We will consider analog input here to limit discussion for this introductory chapter.

Our simple application used only one temperature transducer, but most applications require multiple transducers. To determine the number of channels, we need to determine the transducer types and number of transducers in the target application considering future expansion plans. If all transducers in the application are single ended, we need one DAQ channel per transducer, but if they are differential, we need two channels per transducer.

Choose a device that will sample fast enough for our application and future expansion plans. Our simple application sampled at a low rate since temperature isn't expected to change quickly. We will discuss high-rate applications in Chapter 8. We need to determine the sampling rate that will accurately capture the physical property we are measuring and must consider that our DAQ system will most likely be measuring multiple properties from multiple channels. Commonly, DAQ devices use multiplexing in sampling multiple channels. They sample the first channel, then move to the second, and so forth. They don't sample all channels simultaneously. It is possible to sample several channels simultaneously with multiple boards or Field Programmable Gate Array (FP6A) based hardware.

Evaluate the application to determine the allowable digitizing error. The device must have sufficient ADC resolution to fall within the allowable error. Our temperature-measurement application with the PCI-6024E had a resolution of $+/- 0.244°C$ over a range of outdoor measurements. If our application requires a better temperature resolution say, $0.1°C$, then we must purchase a DAQ device with better resolution. There is an accuracy calculator on ni.com.

Determine the maximum range of the signals from the transducers. We determined that the temperature range of interest in our application was $\pm40°C$ or $80°C$. This gave us a signal output range of 0.8 V ($\Delta V = 80°C/(100°C/V)$). Calculate the output signal range for each transducer in a multitransducer application and select the maximum range for the DAQ device specification. Then, check the digitizing resolution for the selected range.

MODIFYING THE CONFIGURATION

If you want to change the channel configuration to channel 1, for example, you can open the DAQ Assistant configuration window by right-clicking on the DAQ Assistant icon on the Block Diagram or by right-clicking the icon and choosing Properties. Click the change Physical Channel button, as shown in Figure 3.20. Right-click the

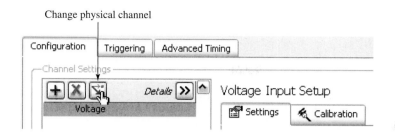

Change physical channel

FIGURE 3.20 Changing channel number in the DAQ Assistant

FIGURE 3.20 Changing channel number in the DAQ Assistant

channel name (Voltage in Figure 3.20) to rename it. Select the Show Details >> Button to show the device and channel configuration.

We can use the program we just developed with different devices without changing the code. If we have a different board instead of the PCI-6024E, we can change the device in MAX and the code will still work.

The DAQ Assistant makes it easy to develop complex data acquisition applications by providing one interface for many types of operations. For example, rather than using a Digital Read function to read data from digital lines and an Analog Read function to read analog data, you use one configuration interface for reading both digital and analog data. The combination of functions into one interface results in a flatter learning curve for you, not only for one device but also for an entire family of devices. Rather than learning four different techniques to program the four types of operations (analog input, analog output, digital I/O, and counter/timer), you need to learn one technique and reuse that knowledge to program the others.

DAQ ASSISTANT CONNECTION DIAGRAM

The temperature transducer in our application is built in to the BNC-2120, but it is common to connect transducers to terminal blocks and cables. For example, you might have a cable from the PCI-6024E to a terminal block like the one shown in Figure 3.21.

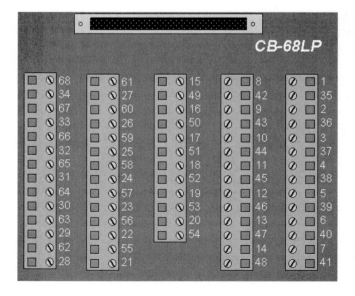

FIGURE 3.21 CB-68LP terminal block (ni.com)

FIGURE 3.22
Replacing the BNC-
2120 with a CB-68LP
terminal block in MAX

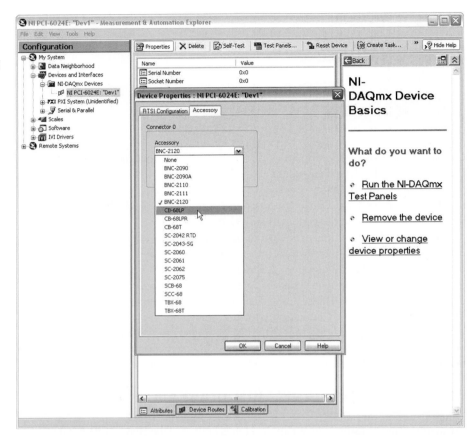

If so, you have to determine the number of the terminal that connects to the channel you configured in the DAQ Assistant, and you have to look up the pinout diagram to determine, for example, what channel corresponds to pin number 68.

So, if we replace the BNC-2120 accessory in MAX with the CB-68LP (as shown in Figure 3.22), the Connection Diagram tool in the DAQ Assistant (Figure 3.23) will tell us which terminals to use.

The Save to HTML option in the Connection Diagram tool will save an HTML report of the task to a file that contains lists of virtual channels in the task, physical channels, device types, measurement types, and connection diagrams.

THE BNC-2120 FUNCTION GENERATOR

The BNC-2120 contains a circuit that generates sine, triangle, and TTL-compatible square waveforms. Use the switch below the Sine/Triangle BNC connector (item 10 in Figure 3.4) to select a sine wave or triangle wave output for the Sine/Triangle BNC connector. A TTL-compatible square wave is always present at the BNC connector labeled TTL Square Wave. To adjust the frequency of the function generator, select the frequency range using the Frequency Selection switch: 100–10 kHz, 1 k–100 kHz, or 13 k–1 MHz (item 11 in Figure 3.4). The Frequency Adjust knob adjusts the

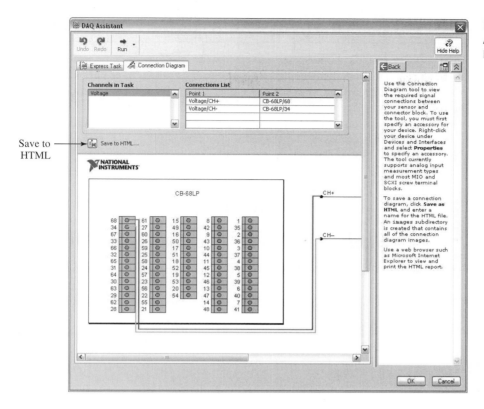

Save to HTML

FIGURE 3.23 DAQ Assistant Connection Diagram tool

frequency within the selected range for the sine, triangle, or the TTL square-wave outputs. The Amplitude Adjust knob adjusts the amplitude of the sine or triangle wave output up to 4.4 Vp-p (items 8 and 9 in Figure 3.4).

THE BNC-2120 QUADRATURE ENCODER

Quadrature encoders measure rotation, for example, in a motor shaft or the joint in a robot manipulator. The BNC-2120 contains a mechanical quadrature encoder circuit that produces 96 pulses per encoder revolution. Two signals, CLK and UP/DN, are available at the screw terminals located below the quadrature encoder knob (items 3 and 4 in Figure 3.4). CLK outputs a pulse train generated by turning the encoder knob with your fingers. It provides four pulses per one mechanical click of the encoder. UP/DN outputs a high or a low signal, indicating rotation direction. If the direction is counterclockwise, UP/DN is low. If the direction is clockwise, UP/DN is high. To use the quadrature encoder, connect CLK to PFI 8, and connect UP/DN to DIO6 (Digital Input/Output Channel 6), which is the up/down pin of counter 0. To use it with counter 1, connect CLK to PFI 3, and connect UP/DN to DIO7, which is the up/down pin of counter 1. We can convert the count from the counter into total degrees as follows:

$$\text{Total Degrees Rotated} = \text{CLK pulses} * 3.75°/\text{pulse}$$

where $3.75°/\text{pulse} = 360°/96 \text{ pulses}$

CODE DEVELOPED BY THE DAQ ASSISTANT EXPRESS VI

The DAQ Assistant developed some code for us. We can view it by right-clicking on the DAQ Assistant and choosing Open Front Panel, as shown in Figure 3.24, which converts the DAQ Assistant to a standard sub VI. Figure 3.25 shows the code built by the DAQ Assistant.

FIGURE 3.24
Converting the DAQ
Assistant to a
standard sub VI

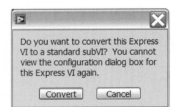

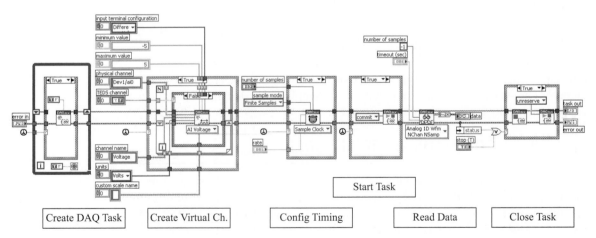

| Create DAQ Task | Create Virtual Ch. | Config Timing | Read Data | Close Task |

FIGURE 3.25 Code built by the DAQ Assistant

SUMMARY

> DAQ automatically enters the data into computer memory and saves it to a file, which is less time-consuming and more accurate than to manually record measurement values.

> Transducers convert physical properties, such as air velocity, into electrical signals.

> A digital (or binary) signal has only two possible discrete levels in amplitude—high level (on) or low level (off). An analog signal, on the other hand, may take on any possible value in amplitude within a specified range.

➤ DAQ devices convert analog signals to digital values and perform some signal conditioning.

➤ Signal conditioning is the process of measuring and manipulating signals to improve accuracy, isolation, filtering, and so on.

➤ The Measurement and Automation Explorer (MAX) program facilitates viewing devices and instruments connected to a computer, and configuring hardware and software.

➤ The DAQ Assistant is a graphical interface (called an express VI) for configuring measurement tasks, channels, and scales.

➤ A physical channel is a hardware terminal or pin at which a signal is connected. A virtual channel is a collection of property settings that can include a name, a physical channel, input terminal connections, the type of measurement or generation, and scaling information.

➤ A NI-DAQmx task is a collection of one or more virtual channels with timing, triggering, and other properties.

➤ Converting analog signals to digital values results in a digitizing resolution error that is also known as measurement precision or quantization error. The error can be reduced by setting the signal range correctly and using high-resolution devices.

➤ A grounded signal source is one in which the voltage signals are referenced to a system ground.

➤ A referenced single-ended measurement system measures voltage with respect to the measurement system ground.

➤ In a nonreferenced single-ended system, all measurements are still made with respect to a single-node analog input sense (AISENSE on E Series devices), but the potential at this node can vary with respect to the measurement system ground.

➤ Differential or floating measurements are not referenced to ground. Differential measurements require two channels.

➤ The output signal from the DAQ Assistant is the dynamic data type. The dynamic data type is for use with Express VIs. The dynamic data type includes attributes that provide information about the signal, such as the name of the signal or the date and time the data was acquired.

➤ We can view the code developed by the DAQ Assistant by right-clicking on the DAQ Assistant and choosing Open Front Panel.

EXERCISES

1. Read this chapter carefully and take notes for your course notebook, including new data types and shortcuts.

2. Build and test the Temperature VI while you are reading. When it is completed, test it by gently placing your finger on the thermistor and observe the change on the graph. If you press down hard, the transducer will be damaged.

3. Research information on the Web to find applications of temperature measurement—for example, in vehicle engine health monitoring or fuel-cell testing. Enter the results in your notebook, including the Web site address.

4. Open the DAQ Assistant. Is your board configured for
   a. differential,
   b. referenced single-ended, or

   c. nonreferenced single-ended measurements?

   _____

5. Change the Temperature VI DAQ Assistant to referenced single ended, and test the temperature VI. What is the result?

   _____

   _____

6. Change the Temperature VI DAQ Assistant to nonreferenced single ended, and test the VI. What is the result? Change the DAQ Assistant back to the correct configuration.

   _____

   _____

7. Modify the sampling information in the Temperature VI DAQ Assistant to 100 samples at 20 Hz. Right-click on the graph plot legend and show points and the line. Then test the VI. Explain how changing the sampling configuration changed the results.

_____

_____

8. Modify the sampling information to Continuous in the Temperature VI DAQ Assistant, and explain the difference with results from the VI from question 2.

_____

_____

_____

_____

9. Modify the sampling information to 1 sample on demand in the Temperature VI DAQ Assistant, and explain the difference with results from the VI from question 2. To see the point, right-click the plot legend (the small window next to the Voltage label above the graph, choose common plots, choose a plot style that displays points). Return the sampling configuration to match that used in question 2.

_____

_____

_____

_____

10. Use a custom scale in the DAQ Assistant to convert to temperature units instead of using the Multiply Function. Change the Range when you add the scale. Name the scale Celsius.

11. Convert to Fahrenheit instead of Celsius with another custom scale in the DAQ Assistant. Name the scale Fahrenheit.

12. Convert to Kelvin instead of Celsius with another custom scale in the DAQ Assistant. Name the scale Kelvin.

13. Modify the program to display a sinusoid signal from the BNC-2120 Function Generator. Use the Function Generator to test the program. Use the BNC Cable and AICH 1 and plot a 100 Hz sine wave on a VI Chart.

a. Change the amplitude of the sinusoid signal to +/−10 V, +/−5 V, +/500 mV, and +/−50 mV, and adjust the scale in the DAQ Assistant to match.
b. Increase the frequency above 100 Hz to see the effect.

14. Use LabVIEW, a text programming language, or a spreadsheet to manually calculate values for a graph such as in Figure 3.14, but for a 4-bit resolution DAQ board. Draw the graph on paper, with LabVIEW or in a spreadsheet program, and compare it with the graph in Figure 3.14. The graph should display a 0–10 V sinusoid along with a simulated analog-to-digital stair-case shaped function generated at 4-bit resolution. How many points do you have to generate to show clearly the stair-step shape?

15. Calculate the digitizing resolution for the PCI-6024E board, and show results and calculations below.
   a. in V for a −10 to 10 V range _____

   _____

   _____

   b. in °C for a −10 to 10 V range _____

   _____

   _____

   c. in V for a −5 to 5 V range _____

   _____

   _____

   d. in °C for a −5 to 5 V range _____

   _____

   _____

16. Change the scale on the graph vertical axis (by double-clicking on the lower and higher values on the vertical axis) so you can see steps in the signal caused by A/D resolution error using a VI that displays a sinusoid signal from the BNC-2120, with the amplitude of the BNC-2120 sinusoid set to 300 mV. (You might have to right-click the graph and uncheck Autoscale Y.) Change the Signal Input Range to the values in Table 3.1, and compare the height of the steps by completing the table below. Compare height of the steps to the A/D resolution-error calculations in the text. Describe what happens when you input a range value that does not match those in Table 3.1. Write an explanation of the results in a word-processor file and save the file.

Signal Input Range	A/D Resolution from Text	Height of Steps in Your Graph	Graph Vertical Axis Range
−10 to +10 V			
−5 to +5 V			
−500 to +500 mV			
−50 to +50 mV			

17. Explain the results from the −50 to 50 mV range:

    _____

    _____

    _____

    _____

18. Research information on the Web to find applications of encoders—for example, in robotics and servo-motor control. Enter the results in your notebook, including the Web site address.

19. Write a program that will display the number of counts from the BNC-2120 Quadrature Encoder. Just display the cumulative number of counts as you turn the encoder no matter which direction, cw or ccwCLK and PFI8 on the BNC-2120. Use the DAQ Assistant defaults of Rising Edge, Count Up, and One Sample on Demand.

20. Enhance the VI from Exercise 19 by calculating the total number of degrees rotated, including both the cw and ccw directions.

21. Enhance the VIs from Exercises 18 and 20 by using the UP/DN signal to increase the count and degrees rotated when you turn the knob cw and decrease it when you turn the knob ccw. Physically wire between the UP/DN and DIO6 connections on the BNC-2120.

Basic DAQ Software Design and Flow Control

INTRODUCTION

We introduced Fundamentals of Software Design and Flow Control and integrated software and hardware for simple data acquisition applications in the previous chapters. This chapter expands the concepts to more sophisticated DAQ by introducing the fundamentals of software design and flow control. Additional capabilities of selection with a Case Structure, repetition with a While Loop, and timing are explained by developing an example that continuously measures temperature. These capabilities give us the ability to build larger applications that will be introduced in Chapter 5 with hierarchical programming and state machines.

OUTLINE

CONTINUOUS TEMPERATURE MEASUREMENT

We will modify the program developed in Chapter 3 to measure temperature continuously as shown in Figure 4.1. The user will be able to choose how long to acquire data, the interval between samples, and the units.

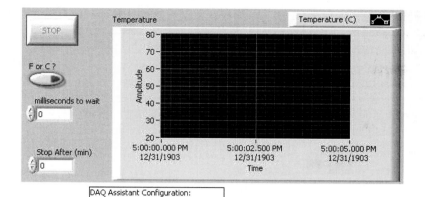

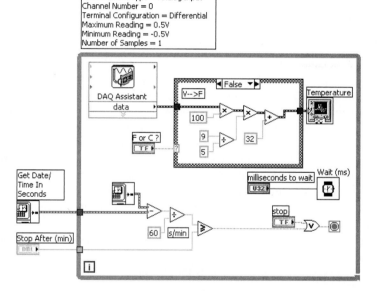

FIGURE 4.1
The Continuous Temperature Measurement Program

The modifications can be quickly completed with the following steps, but we will explain each in detail in the following.

1. Add a Case structure to the Block Diagram.
2. Add a control for acquisition time.
3. Add a control for the sample interval.
4. Code the False Case to convert Voltage to Fahrenheit units and the True Case to convert Voltage to Celsius.
5. Add a Boolean Control for choice of units.
6. Add a While Loop and Stop control.
7. Add time conversion code.
8. Add a greater than or equal to function.
9. Add a wait function.
10. Wire the nodes.
11. Save and run the program.

IMPLEMENTING USER PREFERENCES

This chapter introduces software design. A key element is to understand the needs of whoever is going to use the program. Let's begin with a simple example. Let's assume that the person who is going to use the temperature measurement program wants to be able to change the sample rate, number of samples, and temperature units easily with a GUI when making a series of measurements.

We configured the DAQ Assistant as follows in the previous section:

Measurement Type = Analog Input

Channel Number = 0

Terminal Configuration = Differential

Maximum Reading = 5 V

Minimum Reading = −5 V

Number of Samples = 20

Sample Rate = 2 Hz

This information is "hidden" in the DAQ Assistant Express VI on the block diagram window. In order to put some of this information on the front panel window for user access, we will create controls for the sampling parameters: Sample Rate and Number of Samples. The BNC-2120 requires that we use channel 0 and differential measurement, and our measurement will always be an analog input; so, creating controls for these parameters on the user interface would add unnecessary clutter. However, we can add a free label on the block diagram window to explain the configuration.

We want to control the number of samples and the sample rate. We can do this by adding controls to the front panel window from the Controls Palette as we did in the previous section. But, there is an easier way—creating front-panel controls from block diagram objects. This technique is fast and efficient; because we don't have to worry about matching the data type, we don't have to search the Controls Palette, and wiring is automatic. The first step is to find terminals for the number of samples and the sample rate on the DAQ Assistant. Move the cursor over a terminal on the DAQ Assistant as in Figure 4.2. The cursor changes to the wiring tool, the terminal blinks, and a tip strip appears with the name (label) of the terminal, "number of samples" in this case.

While the terminal is blinking, right-click the mouse to pop up the shortcut menu, shown in Figure 4.3, and choose Create Control.

Repeat this action for the rate terminal and change the label as shown in Figure 4.4. Note that the default value is in parentheses following the name. Align the controls and

FIGURE 4.2

Choosing an input terminal on the DAQ Assistant

Blinking terminal

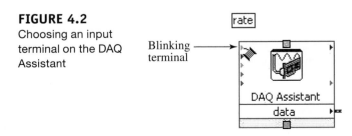

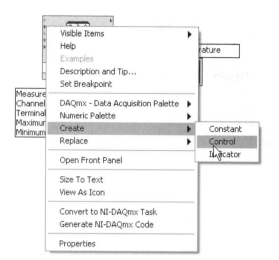

FIGURE 4.3 Creating a Front Panel control from the block diagram window

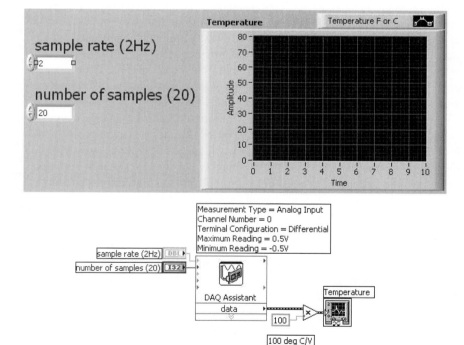

FIGURE 4.4 Sampling parameter controls on the Temperature Measurement Program

labels to complete creating the sampling configuration controls. Save the program. Run the program and test it.

We accomplished something very important by running and testing the program at this point. Since we decomposed our program into steps, we don't have to wait until the program is completely built to test it. We can test increments as we build. By testing increments, we have a better chance that our completed program will be correct. If there is an error, we catch it in a small increment rather than expending the effort to debug a large system to locate the error.

ALGORITHMS, PSEUDO CODE, AND FLOWCHARTS

In the first three chapters, we introduced the concept of implementing automated measurement with data acquisition and graphical programming. We developed some small programs that displayed constants, calculated area, and measured temperature. With this foundation, we can move to more sophisticated and complicated measurement applications.

More-sophisticated applications require planning and design. A good design will make the software and hardware easier to build initially and easier to modify and maintain in the future.

We will begin the design by developing an algorithm. As you probably know from introductory programming courses, an algorithm is a procedure or method for solving a problem or producing a desired result with a computer. The procedure or method contains the actions the computer should execute and the order in which they are executed. The procedure requires us to decompose the problem into small tasks and to define the relationships between the tasks. We might enhance the temperature measurement application developed in the previous section by allowing the user to select units and sampling parameters through front panel controls. To develop a procedure for the application, we might do the problem by hand or with a calculator and write down the steps.

We can write the steps in pseudo code using our native language (English, for example), because it is usually easier to write the steps in our language first and then translate to computer language. That way we divide the design into two steps: define the algorithm, and then translate it into computer language. This technique becomes essential as the problem grows in size and complexity. A complex program may have alternative algorithms, so it is also good to be familiar with the implementation language so we can evaluate them and choose the best alternative.

When designing an algorithm that requires user input, we must think of the possible ways the user could interact and design our code to deal with them. For example, the user might choose either Fahrenheit or Celsius or nothing. If the user doesn't enter anything, we must decide what default value the program should return. The program would be more user friendly if it informed the user which was the default unit.

We used the word "if" in the preceding program description. This implies our code needs to direct the process along different paths depending on the user directives. We can write pseudo code to reflect the alternative paths our code might take to satisfy the user. For example, we might choose to modify the temperature measurement program from Chapter 3 to allow the user to choose units. The pseudo code might be

> Open the program.
> Run the program.
> Read sampling parameters and units from user inputs.
> Acquire a measurement from the temperature transducer signal

> If Celsius units are selected,
>> Convert the units and display temperature in Celsius.
>>> Multiply the voltage by 100.
>>> Display the result.
> Else use Fahrenheit units (use Fahrenheit as default).
>> Convert the units and display the temperature in Fahrenheit units.
>>> Multiply the voltage by 100.
>>> Multiply by 9/5.
>>> Add 32.
>>> Display the result.
> End the program.

We could make Celsius the default, too.

Many software developers represent flow control graphically instead of, or in addition to, using pseudo code. A traditional method is with a flowchart, as shown in Figure 4.5, for the modified Temperature Measurement Program.

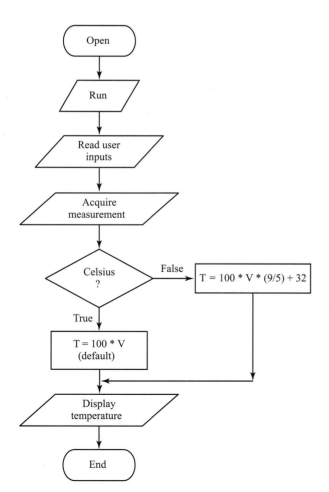

FIGURE 4.5 Flowchart for the Temperature Measurement Program

The flowchart in Figure 4.5 uses four symbols to represent different operations. The oval shape represents program terminations, diamonds represent selections, parallelograms represent input and output, and rectangles represent all processes. A small circle can be used as the beginning and end when showing a subset of the entire chart. Arrows show the flow of control between elements. Some developers prefer to use a larger set of more detailed symbols. (Such a set can be seen in the flowchart drawing-symbol tools in Microsoft Office Word.)

If you draw your program on the computer or scan it to a picture file, you can paste it onto the block diagram window (Figure 4.6). This procedure uses more

FIGURE 4.6
Displaying a flowchart on the Block Diagram

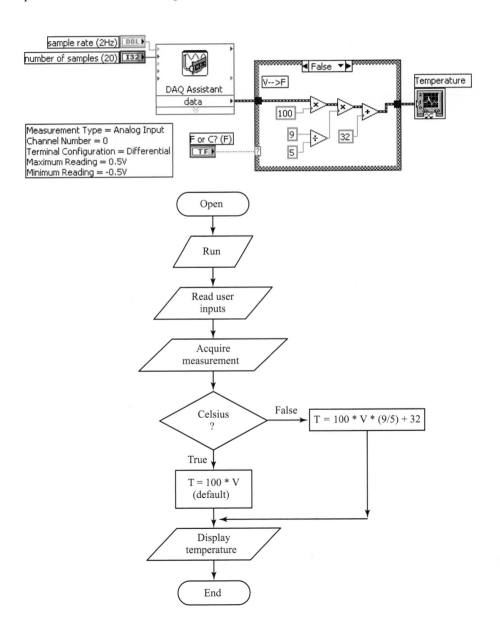

memory and slows the execution slightly, but it will not be a significant factor in the programs developed in this text.

SELECTION WITH A CASE STRUCTURE

Next we will give the user the choice of units using the LabVIEW Case structure to control the flow of the program. If the user selects Celsius, one case (or subdiagram) will implement the Celsius conversion formula, or another case will execute the Fahrenheit formula.

A Case structure in LabVIEW is similar to an "if-else" branching control structure in C:

```
If ( conditional _ expression )
      statement 1;
else
      statement 2;
```

Make some room for the Case structure as shown in Figure 4.7. There are several techniques we can use to make room for the Case structure:

1. Select an individual object to be moved with the cursor, and drag it to the new location. Hold the shift key down to select multiple objects.
2. Draw a window around all of the objects, as shown in Figure 4.7, highlighting the group, and drag the group.
3. Once objects are selected with either of the above two procedures, you can drag them with the mouse or use the arrow keys. If you use the mouse, click on one of the selected objects. Clicking outside of the object will deselect. If you use the arrow keys and want to move long distances, hold the shift key down for larger increments when you press an arrow key.
4. Expand the space between the objects. Hold the Ctrl key down, press the left mouse button, and drag the mouse.

Experiment with these techniques to decide which you like best and become proficient at it so you can add space quickly.

Insert the Case Structure (Functions Palette >> Structures), as shown in Figure 4.8. When we select the Case Structure on the palette, the cursor shape changes to a

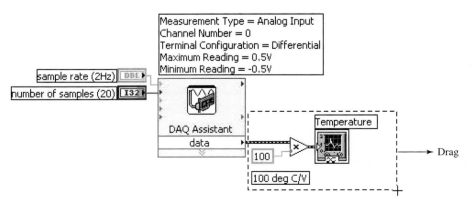

FIGURE 4.7
Creating space for the Case Structure

FIGURE 4.8 Insert a Case Structure from the Functions Palette

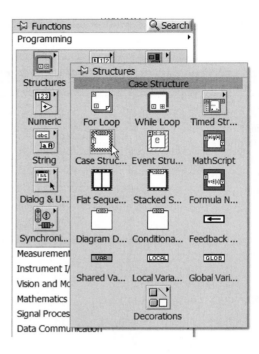

square whose upper-left corner is bold. When we click the left mouse button, it will pin the upper-left corner of the Case Structure at this location. Consequently, before clicking, move the cursor so the upper-left corner is above the existing code, but partway between the DAQ Assistant and the Temperature (C) chart indicator, as shown in Figure 4.9. After pinning the upper-left corner, drag the Case structure lower-right corner to the position shown in Figure 4.10, using the dashed outline to show where the Case structure border will be placed.

Then, click again to pin the lower-right corner, and the dashed outline changes to the Case structure. This sets the size of the Case structure. It is best to make it large enough for the code for all cases, but it can also be resized later. Remember this

FIGURE 4.9
Positioning and sizing
a Case structure

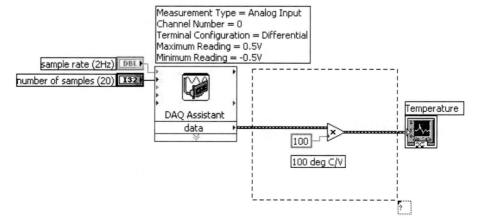

Broken run arrow   Tunnel   Selector label   Tunnel

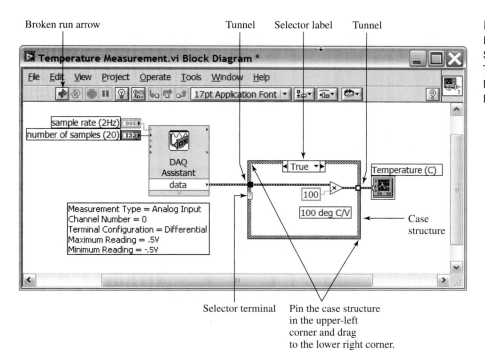

FIGURE 4.10
Inserting the Case
Structure in the
Temperature
Measurement
Program.

Selector terminal   Pin the case structure
in the upper-left
corner and drag
to the lower right corner.

technique. We will use it for other structures, in particular, While Loops and For Loops. After we place the Case structure on the block diagram window, the Run button arrow is broken, as shown in Figure 4.10, to remind us there are additional steps to programming the Case structure.

If we click the Run button, it will display an Error list, shown in Figure 4.11, instead of running the program.

FIGURE 4.11 Error list

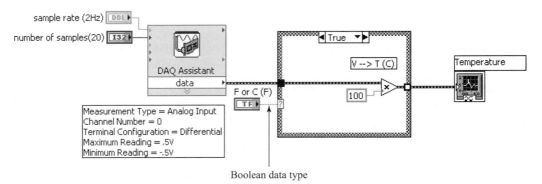

FIGURE 4.12 Adding a Boolean control to the Case Structure Selector Terminal

We will fix the "unwired selector" error first. The green "?" on the left side of the Case structure in Figure 4.10 is the Case structure selector terminal. In the default condition, it accepts a Boolean (True or False) data type control or constant as input. The value wired to the selector terminal determines which one of the Case structure subdiagrams executes, allowing us to implement the decision from the flowchart (Figure 4.5). If the input is true, the program executes the code inside the True case. As we will discuss later, the selector terminal will also accept string, integer, or enumerated data type and allow the program to branch between more than two cases or subdiagrams.

Move the cursor carefully over the selector terminal until the cursor changes to the Select tool shape. Select the terminal and move it lower on the Case structure border, as shown in Figure 4.12. Move the cursor over the Selector Terminal again and right-click. Use the shortcut menu to create a control. Label it F or C? (F) to inform the user that this control changes the units and F is the default. When it is depressed on the right (below the C), the control is true, and the True or Celsius case will execute. If it is pressed on the left or stays in the default position, the False or Fahrenheit Case will execute.

GUIs are easier to use if inputs are on the left and outputs on the right. So, place the controls to the left of the graph indicator. Also, remove the units on the Graph Label since it will display either temperature unit.

Now we will address the "Tunnel: missing assignment to tunnel" error. The True Case is displayed when we first add the Case structure to the block diagram window. The True Case implements the Voltage to Celsius formula, and the data flows out of the structure to the chart. When wires cross the border of a structure such as the Case Structure, they create a tunnel for the data to flow into or out of the structure. This is the source of the other error in the Error list because when we wire out of one case, we must connect the output in the other cases, too. This means we need to switch to the case that implements the Fahrenheit conversion, code the formula, and wire to the tunnel. When the tunnel is wired correctly, the tunnel will fill with color.

Switch from the True to the False Case by clicking on one of the arrows in the Case structure selector label. Complete the coding, as shown in Figure 4.13. The controls represent the Read User Inputs block in the Figure 4.5 flowchart. The DAQ Assistant represents the Acquire Data block in Figure 4.5. The Case Structure represents the decision diamond and associated calculations in Figure 4.5. The Temperature Indicator implements the Display block in the Figure 4.5 flowchart. The Run

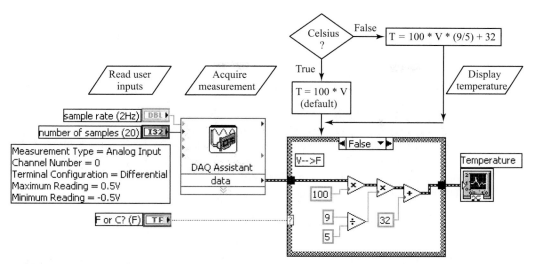

FIGURE 4.13 Temperature Measurement Program False Case

button will turn solid when all output tunnels are wired. Save the program and test it. Note that the Y-axis scale on the graph adjusts automatically when you change units.

We need to change the default value of the Boolean control to match our flow chart and label. We specified that the default conversion was to Celsius which is in the True Subdiagram. However the Boolean control default is False. Change the control to True, right-click the Control, choose Data Operations from the short-cut menu, and choose Make Current Value Default as shown in Figure 4.14.

We click the decrement and increment arrows in the selector label to scroll through the available subdiagrams. We can also add, duplicate, rearrange, or delete the subdiagrams by right-clicking on the selector label and choosing the appropriate

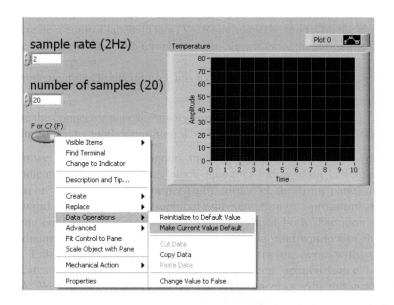

FIGURE 4.14
Changing the Default Value of a Control

operation from the pull-down menu. We can use the labeling tool to enter a single value or lists and ranges of values in the case selector label. We can create multiple input and output tunnels and specify a default case for a Case structure. So, the Case structure has many more capabilities than we need for our temperature measurement program, and we will explore them later.

When data types other than Boolean are used, one of the subdiagrams must be a default in case the input to the Case Structure Selector Terminal doesn't match any of the cases. We can choose which subdiagram to use for the default by right-clicking on the subdiagram label and choosing "make this case default."

We don't have to wire the output tunnels in all cases. LabVIEW gives us the option of using a default value by right-clicking on the tunnel and choosing Use Default if Unwired. But it is better to wire all cases. Wiring all cases makes us think about the program flow, especially when we have numerous cases. Plus, many programmers don't think about what value gets sent out of the tunnel when "Use Default if Unwired" is selected. This leads to some confusing, and difficult to debug, results. So, please don't do it.

Earlier in the chapter, in the section Implementing User Preferences, we showed a text program where variables and their data type were defined before they were used. Placing an icon on the front panel or block diagram window of a LabVIEW graphical program is similar to defining variables and assigning data types. The wires and icons on the block diagram are different colors to make it easy to identify the different data types in our program. Show the Context Help (Help >> Show Context Help, or Ctrl H) and hover the cursor over the wires. Their data type will be displayed in the Context Help window (Figure 4.15).

The Block Diagram in Figure 4.13 and Figure 4.15 contain the following data types:

> double-precision floating point (orange)
> Boolean (green)
> integer (blue)
> dynamic (heavy dark blue with white dashes)

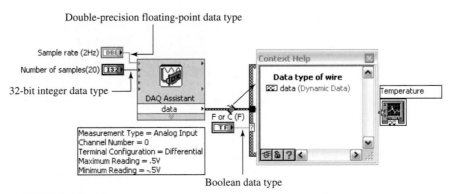

FIGURE 4.15 Using Context Help to discover wire data types

You can see the colors in parentheses in the list of data types if you are working in LabVIEW while reading the text.

Locate these data types in the list of LabVIEW control and indicator data types in Table 2.1.

REPETITION WITH A WHILE LOOP

So far we have used sequence and selection. We need one more form of control to build powerful, complex programs—repetition. There are several repetition structures in LabVIEW, including a While Loop and a For Loop. We will use the While Loop here and introduce the For Loop later.

We will add a While Loop to our Temperature Measurement Program as a matter of instruction and convenience. The loop will allow the user to display a series of values continuously without pushing the Run button each time or without using the Continuous Run button. The user will push a Stop button to end the program.

The addition of the While Loop changes the pseudo code to the following:

> Open the program.
> Run the program.
> Read the sampling parameters and units from user inputs.
> Acquire a measurement from the temperature transducer signal.
> If Celsius units are selected,
>> convert the units and display temperature in Celsius.
>>> Multiply the voltage by 100.
>>> Display the result.
> Else use Fahrenheit units (use Fahrenheit as default).
>> Convert the units and display the temperature in Fahrenheit units.
>>> Multiply the voltage by 100.
>>> Multiply by 9/5.
>>> Add 32.
>>> Display the result.
> If Stop is true, end the program.
> Else return to step 3.

The flowchart changes, as shown in Figure 4.16.

To implement repetition in the program, insert a While Loop from the Functions >> Structures palette, as shown in Figure 4.17. Use the same procedure for inserting it as we did for the Case Structure, pin the upper-left corner and drag the lower-right corner into position. Right-click the While-loop Conditional Terminal and create a control to stop the loop. The While Loop in Figure 4.17 represents the Stop? decision diamond and flow branches in Figure 4.16.

The While Loop repeats the code it encloses until the While-loop conditional terminal, in the lower-right corner of the loop, receives a Boolean value of True. We could change the While-loop conditional terminal to stop when it receives a False value by right-clicking the conditional terminal and selecting Continue if True from the shortcut menu. We will use the Stop if True behavior through this chapter.

FIGURE 4.16

Continuous Temperature
Measurement flowchart

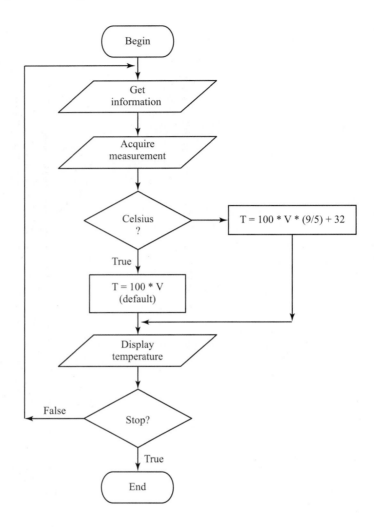

Save the program as Continuous Temperature Measurement. Run the program and test it. Notice that the graph doesn't display anything until all of the samples are acquired, which takes 10 s because the program acquires 20 samples at 2 Hz for each iteration of the While Loop. This is a good example of LabVIEW's data-flow paradigm. The graph doesn't execute until the DAQ Assistant is finished and sends its output to the graph.

Note that the data from the previous loop iteration remains on the graph until it is replaced by the data from the next iteration. We can change the sampling parameters while the program is running, and LabVIEW will read them before acquiring data in each while loop iteration. Change the number of samples and the sample rate and note the result. If we move the controls before the While Loop, the user must restart the program before LabVIEW will read them again, another good example of the data-flow paradigm.

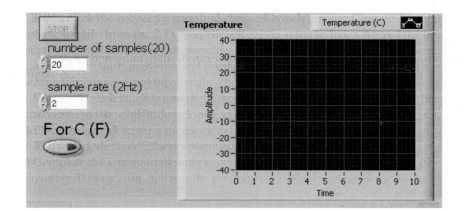

FIGURE 4.17
Repetition with a
While Loop in the
Continuous
Temperature
Measurement
Program

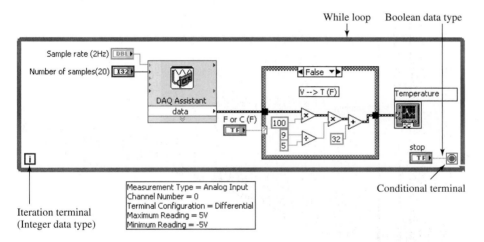

If we move the graph out of the While Loop and pass data to it through a tunnel, it will display data only after we press the Stop button. Again, considering the data-flow paradigm, the graph will only receive data after the While Loop stops.

Rather than waiting for 20 samples to be acquired, converted, and displayed, we can observe the value of each data point as it is acquired and converted. Open the DAQ Assistant and change it to acquire one sample on demand, as shown in Figure 4.18.

We don't need the controls for sample rate and number of samples any longer, so delete them. Save the program as Continuous Temperature Measurement, but don't test it yet.

In order to display previous data, we will change the Graph to a Chart, as shown in Figure 4.19. Right-click on the Graph and select Replace from the shortcut menu. This opens the Controls Palette. Navigate to the Waveform Chart icon in the palette as shown.

A Graph will erase and replot each time the While Loop iterates, but a chart will retain the data from previous iterations. We will discuss graphs in relation to arrays of data in Chapter 6.

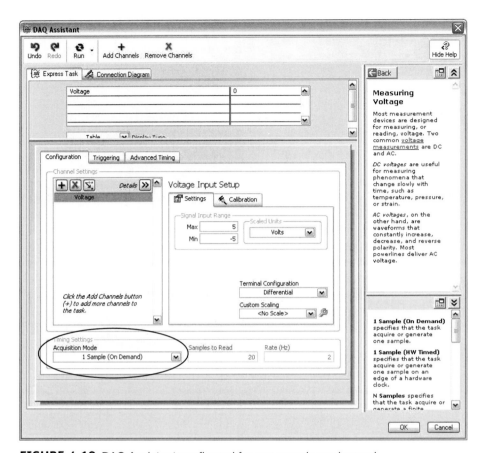

FIGURE 4.18 DAQ Assistant configured for one sample on demand

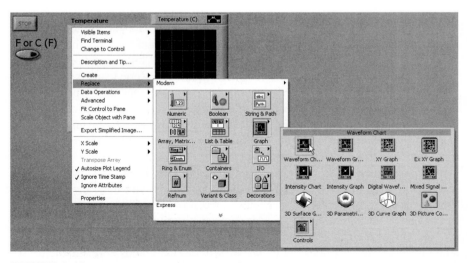

FIGURE 4.19 Replacing the Graph with a Chart

We must always write code that gives the user the option of sending the appropriate value to the conditional terminal that will stop the loop. Otherwise, we create an infinite loop. If this occurs, use the Abort button, or in extreme cases, use the MS Windows Task Manager to end LabVIEW. For this reason, the Run arrow will be broken until the conditional terminal is wired. Even with this safety feature, infinite loops can occur—for example, if the output from a comparison function can never be true and this output is wired to the conditional terminal, say, if we wire 1 to the top of a Greater Than Function and 2 to the bottom. One will never be greater than 2, so the output is never true, but the terminal is wired, so LabVIEW will run.

Do not use the Abort button to stop a program unless it is absolutely necessary. Aborting the program interrupts any process that is underway, and unanticipated events may occur, such as losing data. Always use the Stop control or other appropriate method to end a While Loop gracefully when possible.

The While Loop executes the code in the loop and then checks the value of the conditional terminal input. Consequently, it always iterates at least once. The iteration (i) terminal in the lower-left corner of the loop counts iterations starting at zero for the first iteration. The output-value data-type is integer which will be blue color if you are working in LabVIEW while reading this text.

Save the program and test it. To clear the chart for subsequent runs, right-click on the chart, choose Data Operations from the shortcut menu, and choose Clear Chart.

EXPLICIT EXECUTION TIMING

The While Loop will run as fast as it can and essentially take all of the computer's resources from other important activities. For most applications, we don't need that level of resources or speed, so we will slow the While Loop's iteration frequency by adding the Time Delay VI (Functions >> Timing) inside the While Loop, as shown in Figure 4.20.

Right-click on the block diagram window to open the Functions palette, and navigate to the Time Delay, as shown. When we place the Time Delay VI on the block diagram window, a dialog appears prompting the user to set the amount of time delay. Set it to 0.25 s, as shown in Figure 4.21. This type of timing is called explicit execution timing. It controls how quickly a program executes on the computer processor. It executes the code in the loop and then "sleeps" until the wait time has elapsed.

Save the program and test it. The Time Delay Express VI will accept a front panel control to allow the user to set the amount of time to wait.

Alternately, we can use a Timed While Loop that executes an iteration of the loop at the period specified. It provides multirate timing capabilities, precise timing, feedback on loop execution, timing characteristics that change dynamically, or several levels of execution priority. The timed loop includes (1) input, (2) left data, (3) right data, and (4) output nodes. The developer configures the timed loop by wiring values to the inputs of the input node, or uses the loop configuration dialog box. The left data node of the timed loop provides timing and status information about the previous loop iteration. Refer to LabVIEW Help >> Timed Loop for more information. Right-click the while loop and replace it with a timed loop, as

FIGURE 4.20
Adding a Time Delay
to the Continuous
Temperature
Measurement
Program

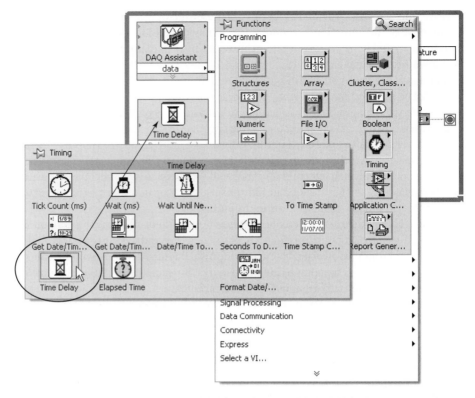

FIGURE 4.21 Setting the
Time Delay

FIGURE 4.22
Timed while loop in
the Continuous
Temperature
Measurement VI

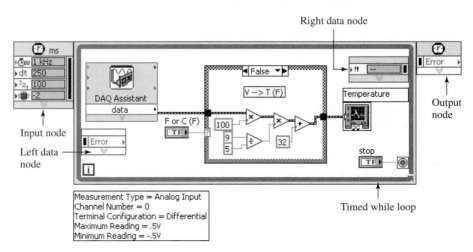

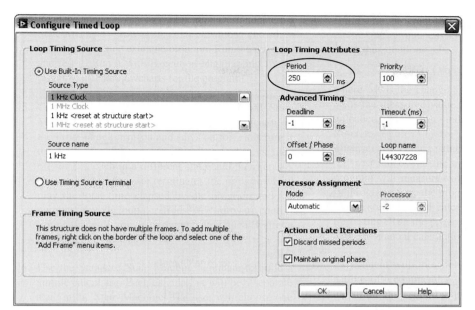

FIGURE 4.23
Timed loop
configuration dialog

shown in Figure 4.22. Delete the Time Delay, then right-click on the border of the while loop and choose replace with timed loop from the shortcut menu. Then double-click on the dt value terminal and change the period to 250 ms in the dialog box (Figure 4.23).

SOFTWARE CONTROL TIMING

We may want to acquire data for a specified time duration. For example, we might want to acquire temperature measurements continuously for 5 minutes. Rather than watching a clock and selecting the Stop button after 5 minutes have elapsed, we can implement code that will automatically stop our program, and the computer will stop it at precisely 5.000 minutes, which would be difficult and tedious for a human to accomplish manually.

Modify the program as shown in Figure 4.24. Add the Get Data/Time in Seconds Function (Functions >> Timing). Make a copy so one function is on the left side of the While Loop so it will output the time when the program starts. Place the copy in the loop to output the time of each iteration. Add a wire to a Subtract Function to calculate the elapsed time.

Create the Stop After front panel numeric control for the user to set the length of time to acquire data. Place a Greater Than or Equal? Function on the block diagram window and wire it to the elapsed time and the Stop-After control to compare the two.

Insert an Or Function in the wire from the Stop button and the While Loop Conditional Terminal. Wire the Stop button and the output from the Greater Than or

FIGURE 4.24
Software control
timing

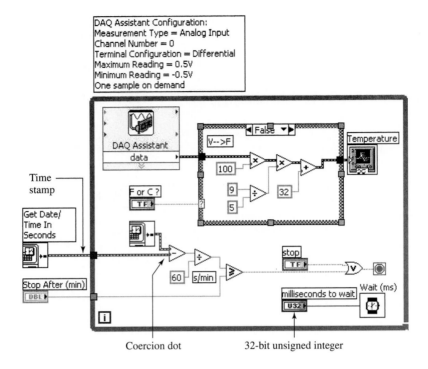

Equal To? Function to the Or Function so the program will stop either when the Stop button is pressed or when the Stop-after time is reached.

Replace the Time Delay Express VI with a Wait Function. Save the program as Continuous Temperature Measurement with Software Timing. Run and test it.

The data type produced by the Get Time in Seconds function is the time stamp data type. It is the number of seconds since 12:00 a.m., Friday, January 1, 1904, Universal Time in 00:00:00 PM MM/DD/YYYY format. LabVIEW automatically coerces it to double-precision floating-point data type at the input terminals to the subtract function. LabVIEW placed a small dot at each location where it coerced a data type. Coercion creates another space in memory for the new data type. If you are programming in LabVIEW while reading this text, you will notice that the dot is red.

Note that since the Stop button is inside the loop, the program checks the value of the Stop button each time the While Loop iterates. If the delay is set to 250 ms, the Stop button is checked approximately every 250 ms. This is important if the user wants to stop the program before the Stop After Time elapses.

The word "approximately" is important in the preceding paragraph. Software timing is not exact. It is not deterministic. The operating system has priority over our program. It can interrupt processing at any time. Therefore, when exact timing is critical, use hardware timing in the DAQ Assistant or use a real-time operating system.

SUMMARY

➤ Design a program before implementing code.

➤ Algorithms are procedures or methods to solve problems with a computer.

➤ Pseudo code expresses an algorithm in a native language (like English).

➤ When designing an algorithm that requires user input, we must think of the possible ways the user could interact and design our code accordingly.

➤ Many software developers represent flow control graphically instead of, or in addition to, using pseudo code. A traditional method is with a flowchart.

➤ In addition to designing the algorithm, we need to design a user interface that is intuitive, so users will understand it easily and enjoy using our program.

➤ The Case Structure implements selection in LabVIEW.

➤ The While Loop repeats code until a termination condition is met.

➤ Explicit execution timing adds a delay to the loop or uses a timed loop to share processor resources with other computer activities.

➤ Software control timing measures elapsed time and performs an action such as stopping the program when the elapsed time expires.

➤ Use hardware timing in the DAQ Assistant or a real-time operating system for more deterministic timing.

EXERCISES

1. Read the chapter carefully and build the Continuous Temperature Measurement with Software Timing program and test it.

2. Update your course notebook from the lecture and reading, including new data types and shortcuts.

3. Place the Stop control outside the While Loop, then run the program and press the Stop button. What happens?

_____

_____

Why?

_____

_____

4. Modify the While Loop conditional terminal in the Continuous Temperature Measurement with Software Timing Program to Continue if True, instead of Stop if True, and change the front panel control accordingly. Test the VI. Change back to Stop if True.

5. Expand the front panel window from the previous question and copy the chart twice. Set one chart to display in sweep mode (right-click the chart and choose Advanced >> Update Mode), the second in scope mode, and the third in strip chart mode. Run the program to observe the difference between chart update modes. Use Stop if True.

6. Modify the Continuous Temperature Measurement with Software Control Timing Program from Figure 4.24 to continuously display a sinusoid signal from the BNC-2120 Function Generator instead of the temperature measurement. Change the time delay to 10 ms get a smoother waveform. Save with a different name.

7. Change the sampling configuration in the VI from the previous question to N samples, 100 samples, and 2 kHz.

   a. How many points does LabVIEW acquire in each iteration of the While Loop?

   _____

   b. Explain how the results compare with the previous question.

   _____

8. Draw a flowchart to design a program that converts a number the user enters in a numeric control from °C to °F or K, depending on which the user selects with a Boolean Toggle Switch control.

9. Implement the program from question 8 in LabVIEW. Save the program as the Temperature Units Conversion.vi. Label the Boolean switch and the indicators appropriately. Display the results on the front panel with a thermometer-shaped indicator. Test the program with known values, such as freezing and boiling at sea level.

---

10. The text didn't cover the Ring control, but it is a good control to use in a program that requires more than two cases. See if you can learn how to use the Ring Control on your own. Write a program that accepts Voltage as an input and converts it to °C, °F, or K, where the user selects the units with a Ring control. Save the program as Voltage Signal to Temperature.vi.

11. Draw a flowchart to design a program that generates random numbers every 0.25 second and turns on a round LED indicator when the random number is less than 0.3. The VI should automatically stop when the random number is less than 0.05 or if 2 minutes have elapsed. Show points where you can incrementally test the program on the flowchart.

12. Implement the program from question 11 in LabVIEW. Save the program as the Random.vi. Run and test the program. Add free labels to the block diagram window that identify all of the data types used in the program.

13. Draw a flowchart to design a program that calculates the square root of a number entered on the front panel window and displays the result on the front panel window. Test the number entered to see if it is negative. Display −99999 in the square root indicator if a negative number was entered, and send a string message to the user requesting a positive number. Show points where you can incrementally test the program on the flowchart.

14. Implement the program from question 13 in LabVIEW. Save the program as the Square Root.vi. Run and test the program. Make a table of the data types used in the program.

15. Draw a flowchart to design a program that outputs the area of a circle when the user inputs the radius. Check for negative input values and have your program react to them in a user-friendly manner. You may modify the program from the previous section, Circle Area VI, developed in Chapter 2. Show points where you can incrementally test the program on the flowchart.

16. Implement the program from question 15 in LabVIEW. Save the program as the Circle Area.vi. Run and test the program. Make a table of the data types used in the program.

17. Write a comparison (in a word-processing program) between the LabVIEW Circle Area Program from the preceding exercise and the following program written in C. Use correct technical writing skills. Which is easier to implement? Justify your conclusion/opinion.

```
/*
 *==========================================================================
 * Compute Area of a Circle: Coefficients are defined in main program.
 *                         : Radius and area of circle are printed to
 *                           screen.
 *
 * Written by: Mark Austin                              January, 1994
 *==========================================================================
 */
#include <studio.h> /* Standard Input/Output function declarations    */
#include <math.h> /* Math functions, such as sqrt(^), and constant M_PI */
```

```
int main( void ) {
float fRadius; /* Radius of circle */
float fArea /* Area of circle     */
float fPi /* Variable for "pi"    */
   /* [a] : Prompt User for "radius of circle"" */

   printf("=======================================================\n");
   printf("Please input the circle radius (Radius > 0):");
   scanf("%f", &fRadius);
/* [b] : Check that the radius is greater than zero. */

  if( fRadius <= 0 ){
    printf("ERROR >> Circle radius must be greater than zero\n");
    exit (1);
  }

  /* [c] : Compute Area of Circle */

  fPi = 4.0*atan( 1.0 );
  fArea = fPi*fRadius*fRadius;

  /* [d] : Print Radius and Area */

  printf("Radius of Circle = %8.3f \n", fRadius);
  printf("Area of Circle = %8.3f \n", fArea);
  return (0);
  }
```

18. Draw a flowchart to design a program that outputs the diameter, area, and perimeter of a circle when the user inputs the radius. You may modify the circle area program from the preceding exercise. Check for negative input values and have your program react to them in a user-friendly manner. Align and organize the indicators on the front panel window to make a user-friendly interface. Show points where you can incrementally test the program on the flowchart.

19. Implement the program from question 18 in Lab-VIEW. Save the program as the Circle Geometry.vi. Run and test the program. Add free labels to the block diagram window to identify all of the data types used in the program.

20. Draw a flowchart to design a program that calculates the length and slope of a line when the user enters two points in 2D space. A point will have an X coordinate and a Y coordinate. Check for divide by 0. Test the program with a vertical line, a horizontal line, a line of length 0, and a line with a slope of 1 and length of 1 in all four quadrants. Display the slope with a dial indicator. Show points where you can incrementally test the program on the flowchart.

21. Implement the program from question 20 in Lab-VIEW. Save the program as the Line.vi. Run and test the program. Make a table of the data types used in the program.

22. Calculate and display the slope and time elapsed between temperature measurement points by adding or copying and pasting code to the Continuous Temperature Measurement with Software Timing Program.

23. Challenge: Display the equation of the line in mx + b format on the front panel window of the line.vi

24. Find a LabVIEW Example Program that calculates a running average. Run it with Execution Highlighting and explain how it works. Make a table of the data types used. Make a table of the constants, functions, indicators, controls, and structures used. Draw a flowchart of its execution sequence, branching, and repetition.

25. Write a program that displays the signal from the BNC 2120 encoder and indicates whether the encoder is turning clockwise or counterclockwise with a round LED.

26. Develop a flow chart for the continuous temperature measurement program with software control timing shown in Figure 4.24.

DAQ State Machines

INTRODUCTION

The previous chapters built small DAQ applications and introduced structures that can be used to build larger applications. As applications increase in size, they become difficult to program correctly. This chapter explains how to decompose large applications into small chunks with hierarchical and state-machine programming. It examines the data-flow paradigm that separates graphical programming from text programming, and uses the Execution Highlighting tool for visualizing data flow and debugging. An example DAQ application is developed with a state machine that incorporates enumerated types, Shift Registers, sub VIs, and more GUI development. Automated DAQ applications typically produce large amounts of data. In Chapter 6, we will introduce the array data structure to manage and manipulate large data sets.

OUTLINE

PROGRAM STRUCTURE

As measurement applications become more sophisticated, we will integrate the flow control techniques defined in Chapter 4: sequence, selection, and repetition. The programs we developed in Chapters 2 through 4 fall in the categories of simple and general architectures, as shown in Figure 5.1 and Figure 5.2. The program that uses simple architecture does not have any flow control structures and runs only once. The General VI architecture has flow control structures, such as loops, with a start-up area before the structure; a main area in the structure; and a shut-down area after the structure. (The Simple Error Handler VI in the shutdown region will be explained later.)

We combined two structures in the simple Continuous Temperature Measurement Program: a Case structure and a While Loop. As our measurement applications become more sophisticated, we will use even more structures. Our options for connecting them include:

> Series or sequential—one structure completes before the next executes
> Nested—one structure executes inside of another (like the Case structure inside the loop—in the Continuous Temperature Measurement Program)
> Parallel—running simultaneously (or sharing multiple threads) with or without connections

We can insert multiple, sequential Case structures, as shown in Figure 5.3, to build a Multiple Case structure architecture. The user chooses which action to execute with front panel controls. Data can be easily exchanged between the Case structures. However, as the number of actions grows, the block diagram stretches beyond the edges of the monitor. One of our objectives is to build readable code. Readable block

FIGURE 5.1
Simple VI
architecture

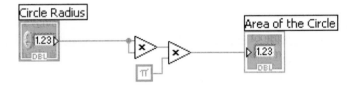

FIGURE 5.2
General VI
architecture

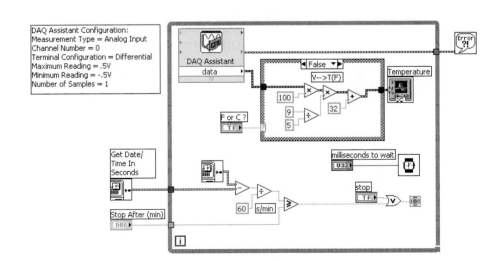

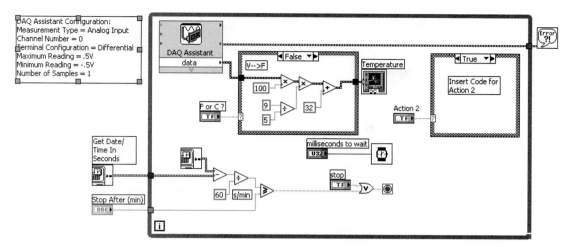

FIGURE 5.3 Multiple Case Structure architecture

diagram windows fit within the monitor, so it is easy to see the entire program. Readability is difficult to accomplish with the Multiple Case Structure architecture. In addition, some events are blocked when others execute, and all events must execute at the same rate.

The State Machine architecture shown in Figure 5.4 is better. It nests a single Case structure in a While Loop. The program in Figure 5.4 is a LabVIEW template, available from the Getting Started Dialog >> VI Template >> Frameworks >> Design Patterns as shown in Figure 5.5. We will cover this architecture in detail in this chapter.

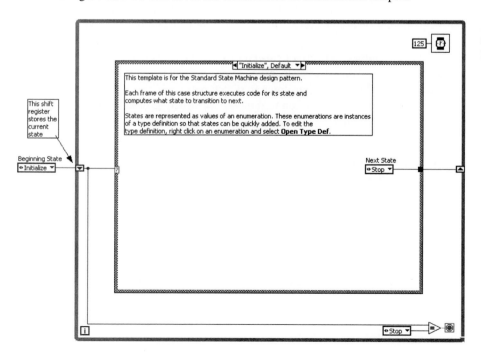

FIGURE 5.4
Standard State
Machine template

FIGURE 5.5
Creating programs
from LabVIEW
templates

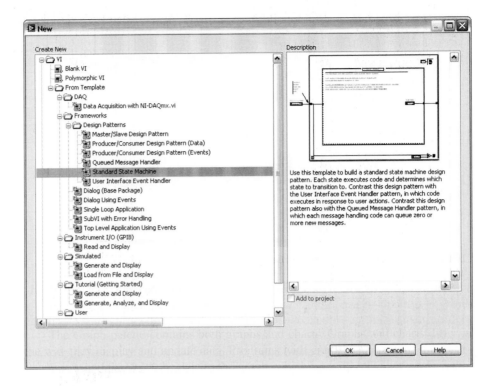

We could also add a separate loop to our application, as shown in Figure 5.6, to build a parallel loop architecture. This architecture is useful because we can run the loops at different rates. We divide a large application into multiple loops so processes that need to execute quickly aren't slowed by other processes. We have to use special variables to share data between parallel loops, which we will discuss later using Master-Slave and Producer-Consumer architectures.

We must pay careful attention to the way we structure or architect our program to avoid writing code that

> we can't reuse and we have to rewrite for new applications,
> is difficult for others to read,
> is difficult to debug,
> has to be rewritten if we want to modify it or,
> takes a lot of effort to test and prove correct.

To demonstrate some effective graphical programming architectures we will use structured programming to build a small program in this chapter. Our problem will be to develop a program that acquires, graphs, and saves temperature data. The goals for our structured programming are:

> Readable
> Scalable
> Maintainable

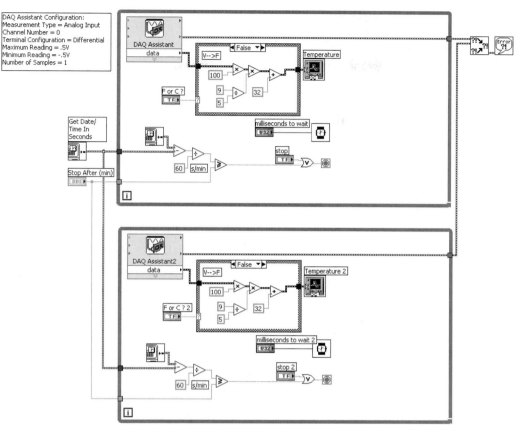

FIGURE 5.6 Parallel loop architecture

The program will be reusable in a larger program, which might measure additional channels, analyze the data, and display both acquired and analyzed data. Our program will follow the user interface and block diagram design rules presented in Chapter 4. We will begin by developing high-level pseudo code and a flowchart (Figure 5.7).

1. Run the program.
2. Acquire 1 point of temperature data.
3. Display the result to the user in a graph.
4. Save the data.
5. If Stop is True, end the program.
6. Else, repeat the acquisition.

This is enough detail to implement our code, but more sophisticated programs might require an additional phase of development via the top-down stepwise refinement process. In top-down stepwise refinement, we expand each of the processes in the high-level flowchart and pseudo code.

FIGURE 5.7 High-level flowchart
for the Temperature Measurement
Program

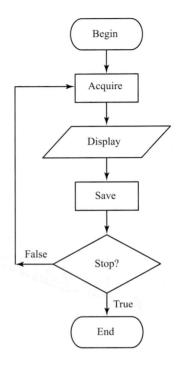

INTRODUCTION TO STATE MACHINES

LabVIEW state machines are based on a "Moore machine," which performs a specific action or task for each state in the diagram. State machines are an easy and elegant architecture for decomposing complex problems and implementing complex decision-making algorithms. Each state can lead to one or subsequent states and can also end the process flow. A state machine relies on either user input or programmatic evaluation to determine which state to execute next. Programs rarely remain static over time. We usually modify them by deleting parts of code and adding additional code. We can add and subtract states easily with a state machine, while maintaining compact code that fits within the monitor screen. Therefore, state machines meet our design goals. They are readable, scalable, and maintainable.

In this chapter, we will learn how to build a program that has a separate state for each task. Our program will execute the states in sequence, but state machines can skip states and runs states multiple times before moving on. We will introduce the state diagram and state tables that help clarify some of these more-complicated flow patterns.

ENUMERATED TYPES

Enumerated types are not essential to all state machines, but they make the code self-documenting and when combined with type-defined controls (which we will explain shortly), enhance scalability. An enumerated type control, or enum, associates numeric values with strings. Enums display the string to the user to improve program readability, but their actual values are 16-bit unsigned integers (U16).

Enumerated-type data takes advantage of the polymorphism of Case structures. Polymorphism is the ability of an object to take different forms. Different data types can be wired to the Case structure selector terminal. We wired a Boolean data type previously, but it will also accept integer (and the enumerated type), string, and error cluster types. We don't have to do anything special to the Case structure to get it to accept the other types—we just have to wire the type into the selector terminal and the Case structure automatically morphs into the new form.

To show this polymorphic capability, place an enumerated control (Controls >> Ring & Enum) on the front panel window of a new VI, as shown in Figure 5.8. If you are programming in LabVIEW while reading this text, you will notice that the block diagram icon is the same blue color as the integer data type.

Right-click the enumerated control and choose Properties. Choose Edit Items, and type State 1, State 2, and State 3, as shown in Figure 5.9.

Add a Case structure to the block diagram, as shown in Figure 5.10, and wire the Enum control to the Case selector terminal. If you are programming in LabVIEW while reading this text, you will notice that the terminal changes color from the default Boolean green to the integer blue, and the case labels change from True and False to State 1, and State 2. The Case structure morphs from Boolean to enumerated and is no longer limited to the two True and False cases.

Because the Case structure default is Boolean, it has only two subdiagrams when we place it on the Block Diagram. Our enumerated constant has three items. If we want to see the third item, we need to add a subdiagram. Right-click on the border of the Case structure and choose "Add case for Every Value." This will expand the Case structure to match the values in the enumerated constant. Click the decrement and increment arrows on the case selector label to view the three different labels.

FIGURE 5.8 Enumerated control

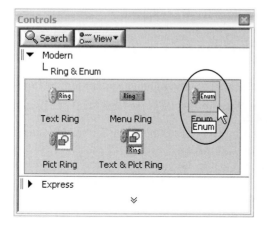

FIGURE 5.9
Adding items to an
enumerated control

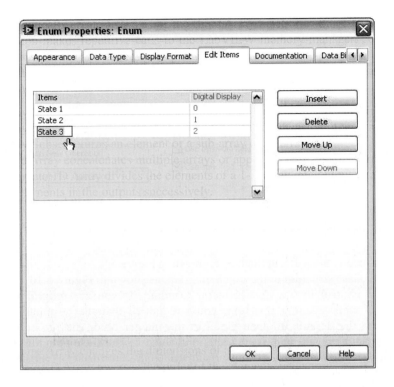

FIGURE 5.10 Connecting
an enumerated control to a
Case structure

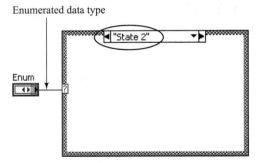

TEMPERATURE MEASUREMENT STATE MACHINE

We will introduce state-machine programming by developing a small program simi-
lar to the Continuous Temperature Measurement VI from Chapter 4. We will start
with the Standard State Machine template (File >> New >> VI >> From Template >>
Frameworks >> Design Pattern), shown in Figure 5.4. Open the VI and save it as
Temperature Measurement State Machine. You will be asked to save additional files.
Save them as well. We will describe them in the following.

The template uses an enumerated constant on the left side of the While Loop.
Click on the down arrow of the enumerated constant to show its two values: Initialize
and Stop, as shown in Figure 5.11.

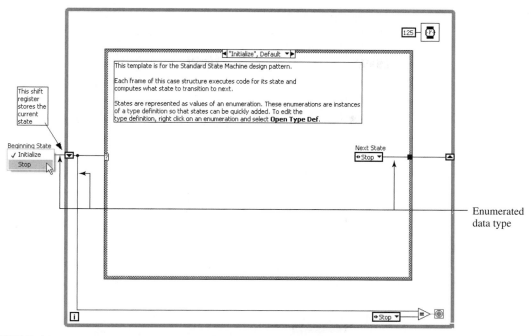

FIGURE 5.11 Standard State Machine template Block Diagram

We need to modify these values by replacing them with the states for our temperature measurement application: acquire, display, and save. The instructions in the free label state, "These enumerations are instances of a type definition so that states can be quickly added. To edit the type definition, right click on an enumeration and select Open Type Def."

A type definition is a master copy of a control or indicator. These constants were created from an enumerated control similar to the process we used previously; however, the Standard State Machine template used a custom, type-defined control. A type-defined custom control is very useful in state machine programming because it will automatically update the constants when we add a new state.

We need to open this control to modify the constant. Right-click on the constant and choose Open Type Def. from the shortcut menu, which opens the custom control shown in Figure 5.12.

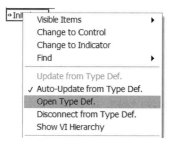

FIGURE 5.12 Open the type-defined custom control

FIGURE 5.13
Edit items in the
type-defined
custom control

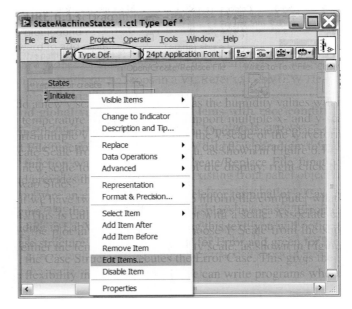

FIGURE 5.13
Edit items in the
type-defined
custom control

The Open Type Def. selection opens the control window shown in Figure 5.13. Right-click on the control and choose Edit Items from the shortcut menu.

Edit the items in the list by replacing Initialize and Stop with Acquire, Display, and Save, as shown in Figure 5.14.

After modifying the names of the states, save the control as the Measurement States.ctl. To apply the changes to the template, select File >> Apply Changes, then

FIGURE 5.14
Replace Initialize and
Stop Item names

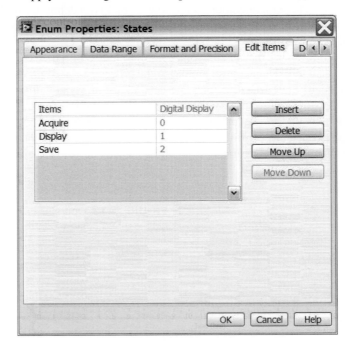

TABLE 5.1 **Simplified State Transition Table**

State	Next state (constant value)
Acquire	Display
Display	Save
Save	Acquire

File >> Close. LabVIEW automatically adds a .ctl extension to identify this file as a control file. Programs with custom controls will not run without this .ctl file. We can use our new control in many VIs, not just this one. Refer to Help >> Index >> Custom Controls for more information about custom controls.

If we lose the file, we can right-click the disabled instance and select Disconnect From Type Def from the shortcut menu. Disconnecting from the type definition removes the restrictions on the data type of the instance, making it an ordinary control, indicator, or constant. If we later find the file, we can replace the control or indicator with the type definition.

Right-click on the border of the Case Structure and choose "Add Case for Every Value." This will expand the Case Structure to three subdiagrams from two (the default). Navigate to the new case, right-click on the Case Structure tunnel, and create a constant for the new case.

Change the values of the constants in each subdiagram appropriately per the simplified state transition table (Table 5.1). The tunnel in the right side of the Case Structure should fill solid with blue, and the Run arrow should be intact.

Save the template as Temperature Measurement State Machine.

We can test the program using Highlight Execution. The program will change to a different task as it runs. We can see the data flow from the enumerated constants to the Case Structure and see the subdiagram switch.

Since the Case Structure inherited labels from an enumerated type, we did not have to type the labels. Notice that the values display in quotation marks, for example, "acquire" and "display." We do not need to type the quotation marks unless the string or enumerated value contains a comma or range symbol ("," or ".."). We use special backslash codes for non-alphanumeric characters, such as \r for a carriage return, \n for a line feed, and \t for a tab. Refer to the LabVIEW Help for a list of backslash codes. When we are programming in LabVIEW, and we enter a selector value that is not the same type as the object wired to the selector terminal, the value appears red to indicate the error.

SHIFT REGISTERS

The enumerated constant is wired to a shift register on the left border of the While Loop. The flow of the program will automatically change from acquire to display, then to save and finally back to acquire. The template implements this by storing the

FIGURE 5.15 Stacking shift
register elements

state name in a computer memory register, called a shift register, which is read by the case selector. After the case selector reads the shift register, the program flow changes to the new state name. The program will check the memory each time the While Loop iterates to see what state to execute next.

Registers are a special, high-speed storage area within the CPU. For example, if two numbers are to be multiplied, both numbers must first be stored in registers, and the result is also placed in a register. Computers usually pass information from memory to a register, operate on it, and pass it back to memory. Computers generally have several different types of registers for storing a variety of types of data such as floating-point registers, constant registers, or address registers. Shift registers shift bits to the left or right. But in LabVIEW, shift registers are used to save data in memory so it is accessible from one iteration of a loop to the next. Conceptually, they can "shift" data between loop iterations. If we refer to the flowchart in Figure 5.7, the value in the shift register needs to begin at Acquire, then Display, then Save, and return to Acquire if the Stop button wasn't pressed.

Shift registers can have more than one element, as shown in Figure 5.15. Multiple elements are added by right-clicking on the shift register and adding elements. The elements access data from previous loop iterations. Consequently, we can use multiple-element shift registers for running average calculations.

The user doesn't need the Measurement States control on the front panel window as the states will change automatically, so we will hide it to reduce clutter on the Front Panel. Right-click on it and choose Advanced >> Hide Control.

Expand the program to create space in the subdiagrams for the new code. Add code to the Acquire state as shown in Figure 5.16.

When you wire the dynamic data type out of the DAQ Assistant to the Case structure border, a tunnel is automatically created. When you wire from the tunnel to the While Loop right border, another tunnel is automatically created. However, we need a shift register instead of a tunnel here. So, right-click on the tunnel and choose Replace with Shift Register. The reason we need a shift register is that we want to write the data from the DAQ Assistant into memory so we can access the memory from other states to display the data on a chart and save the data to a file.

Remember that all output tunnels Case structure must be wired. Wire across the data shift register in all of the states so the Case structure tunnels will fill with color. Save the program.

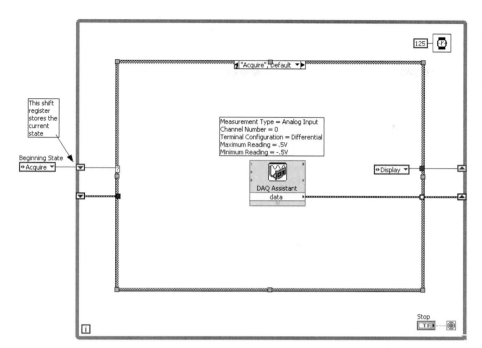

Since shift registers represent computer memory, it is a good idea to define their initial value so the program will start with known values. The signal output from the DAQ Assistant initializes the data shift register in the acquire state. We can also create an array constant to initialize the shift register, which we will do in a later chapter.

Now code the Display state as shown in Figure 5.17.

We would like to test the program before coding the Save state, but the Run arrow is broken. We need to wire the dynamic data tunnel in the Save state before we can test. After wiring to the Data and State Transition Tunnels on the right border of the Case Structure, the tunnels will be filled solid with the color appropriate to their data type. When the tunnels have been wired from all cases, the Run button will be solid. Prior to wiring these tunnels, it will be broken because output tunnels in Case Structures must be wired unless the developer chooses the "Use Default if Unwired" option, which I don't recommend. This option creates problems as it doesn't make developers think about wiring and whether or not the default value is appropriate. Consequently, default values are sometimes passed inappropriately, creating difficult debugging situations.

Complete the wiring as shown in Figure 5.18, and test the program. State Machines allow you to incrementally test programs, which facilitates debugging complex applications.

After testing the Acquire and Display states, we will code the Save state. Change to the Save case and insert the Write to Measurement File Express VI (Functions >> File I/O), as shown in Figure 5.19.

Configure the Express VI, as shown in Figure 5.20. The folder structure on your machine may be different than the one shown, so use a path in the File Name

FIGURE 5.17
The Display state

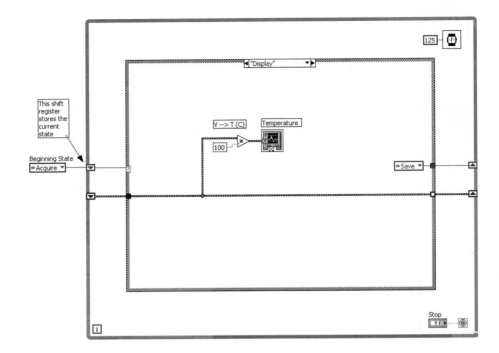

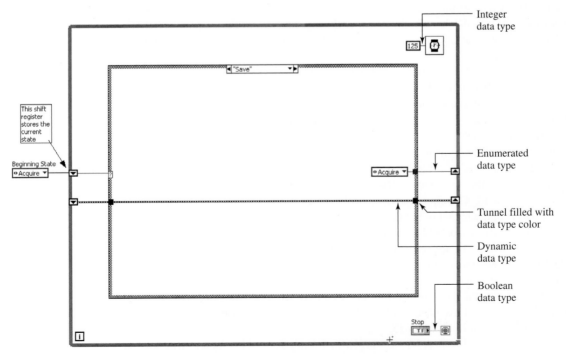

FIGURE 5.18 Wire the dynamic data wire to the tunnel in the Save state

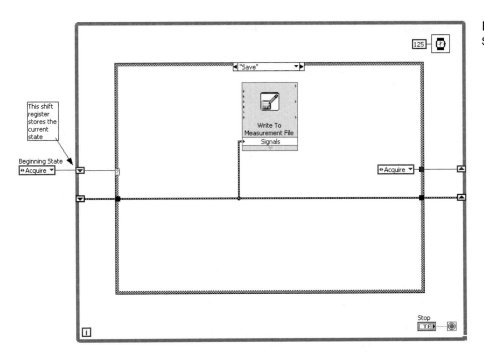

FIGURE 5.19
Save state

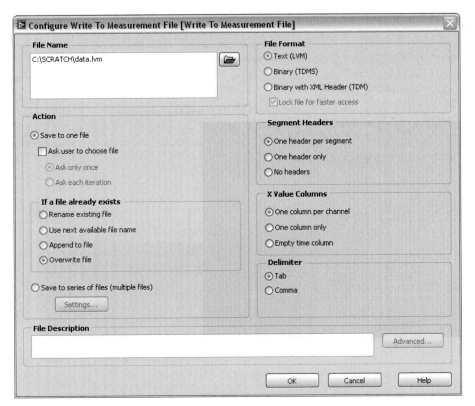

FIGURE 5.20 Write to Measurement File configuration dialog

box that is appropriate for your computer. This VI writes data to a special file format called a LabVIEW Measurement File, which is a tab-delimited text file that we can open with a spreadsheet application, such as Microsoft Office Excel, or a text-editing application, such as Microsoft Notepad. The .lvm file includes information about the data, such as the date and time the data was generated, in addition to the data.

Wire to the Write to Measurement File Express VI. Save and test the program. Open and view the saved file in a spreadsheet application.

The Temperature Measurement State Machine Program uses the following data types. If you are programming in LabVIEW while reading this text, you will notice the colors for each data type shown in parentheses.

dynamic (dark blue with white dashes)

double-precision floating point (orange)

unsigned 16-bit integer (blue)

enumerated (blue)

Boolean (green)

For additional information on state machines in LabVIEW, refer to "Application Design Patterns: State Machines" by Kevin Hogan on www.ni.com.

STYLE

Now that we have a functioning application, we will enhance it by making it easier to use, more readable, and easier to modify. Modify the front panel window as shown in Figure 5.21.

Expanding the size of the front panel window and expanding the size of the chart allows us to view more data. Placing the input controls on the left sends a message that they should be used first. Naming the plot Temperature provides more specific information to a user.

FIGURE 5.21
Graphical user interface for the Temperature Measurement State Machine

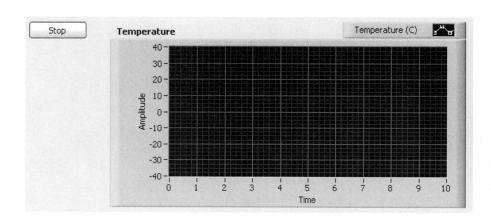

Two lists of programming style-rules follow this paragraph. You may not understand all of them at this point. Follow the ones you understand, and add additional ones as additional material is explained. Refer to the LabVIEW Style Checklist in LabVIEW Help for more information on the items in this list.

Front Panel Rules

➢ Give controls meaningful labels and captions.

➢ Use consistent capitalization, and include default values and unit information in label names.

➢ The Context Help window displays labels as part of the connector pane. If the default value is essential information, place the value in parentheses next to the name in the label. Include the units of the value if applicable.

➢ Set controls with reasonable default values.

➢ Use standard, consistent fonts—application, system, and dialog—throughout all front panel windows.

➢ Use Size to Text for all text for portability, and add carriage returns if necessary.

➢ Use path controls instead of string controls to specify the location of files or directories.

➢ Configure numeric inputs with data ranges.

➢ Write descriptions and create tip strips for controls and indicators, including array, cluster, and refnum elements.

➢ Arrange controls logically.

➢ Group and arrange controls attractively.

➢ Use color logically, sparingly, and consistently, if at all. Use a minimal number of colors, emphasizing black, white, and gray. Use light gray, white, or pastel colors for backgrounds. The first row of colors in the color picker contains less-harsh colors suitable for front panel window backgrounds and normal controls. The second row of colors in the color picker contains brighter colors you can use to highlight important controls. Select bright, highlighting colors only when the item is important, such as an error notification.

➢ Provide a Stop button if necessary. Do not use the Abort button to stop a VI. Hide the Abort button.

➢ Use ring controls and enumerated type controls where appropriate. If you are using a Boolean control for two options, consider using an enumerated type control instead to allow for future expansion of options.

➢ Make sure all the controls on the front panel window are of the same style. For example, do not use both classic and modern controls on the same front panel.

➢ Use imported graphics to enhance the front panel window. One disadvantage of using imported graphics is that they slow screen updates.

➢ Configure the front panel window to fit on the screens of most users.

Block Diagram Rules

Avoid creating extremely large block diagram windows. Block diagram windows should fit within the dimensions of the monitor. If you must have a larger diagram, limit the scrolling necessary to see the entire block diagram window to one direction.

Developers who maintain and modify VIs need good documentation on the block diagram. Without it, modifying the code is more time-consuming and error prone.

Use free labels on the block diagram to explain what the code is doing.

Use the standard application font in free labels on all block diagrams.

Use Size to Text for all text for portability, and add carriage returns if necessary.

Make sure data flows from left to right and wires enter from the left and exit to the right.

Use small free labels with white backgrounds to label long wires to identify their use.

Add as few bends in the wires as possible and keep the wires short. Avoid creating wires with long complicated paths because long wires are confusing to follow.

Align and distribute functions, terminals, and constants.

Avoid placing block diagram objects, such as subVIs or structures, on top of wires, and do not wire behind objects.

Use path constants instead of string constants to specify the location of files or directories.

Make sure the program can deal with error conditions and invalid values.

Use sequence structures sparingly because they hide code. If flow-through parameters are not available and you must use a sequence structure in the VI, consider using a Flat Sequence structure.

If you open references to a LabVIEW object, such as an application, control, or VI, close the references by using the Close Reference function. It is good practice to close any reference you open programmatically.

Avoid using local variables when you can use a wire to transfer data. Every local variable that reads the data makes a copy of the data. Use global and local variables as sparingly as possible.

MODULAR PROGRAMMING WITH SUB VIs

A subprogram, or sub VI, is any program that is imbedded in or called from the block diagram window of another program. Subprograms create a hierarchical structure that supports a good programming practice called modular programming. It allows us to divide complex problems into smaller, more manageable, blocks or modules. We can implement each module as a subprogram, test it, and assemble the modules hierarchically into a large, sophisticated project. When we begin with high-level pseudo code and refine it using the stepwise top-down approach, we are building a hierarchy that can be implemented with subprograms. Some other advantages of modular programming:

Subprograms support incremental testing, because each can be tested individually before it is added to a project.

Subprograms support teamwork, as team members can individually build modules that are assembled into a project.

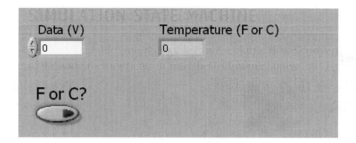

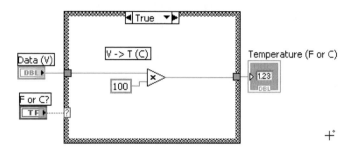

FIGURE 5.22

Temperature Units
Conversion VI

Subprograms support code reuse, as we can use them in other programs with minimal modifications if they are designed properly.

To demonstrate modular programming, we will put the units conversion code into a subprogram. To begin, open a new VI and copy the conversion code into it as shown in Figure 5.22. Right-click the data tunnel on the left border of the Case Structure and create a control labeled Data (V). Right-click the tunnel on the right border and create an indicator labeled Temperature, as shown. Save the program as Temperature Units Conversion.

To implement the Temperature Units Conversion as a subprogram, we

1. create an icon,
2. define the connector pane terminals,
3. and insert it into another program.

When we run the calling program, the subprogram controls receive data from, and the indicators return data to, the block diagram window of the calling program.

The icon is located in the upper-right corner of the front panel and block diagram windows. An icon is a graphical representation of a program's functionality. It can contain text, images, or a combination of both. The icon identifies and self-documents the subprogram on the block diagram window of the calling program. LabVIEW provides a default graphic for the icon with a number that indicates how many new programs have been opened since launching LabVIEW. We can double-click the icon to customize or edit the graphic so it

FIGURE 5.23
Icon Editor

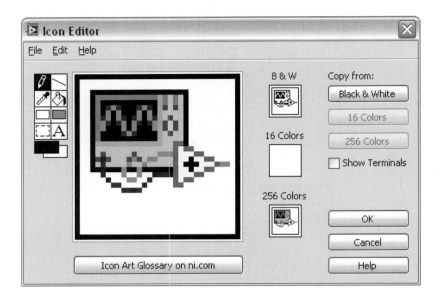

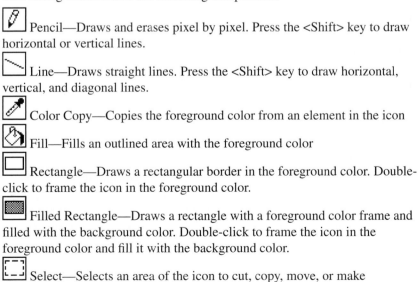

represents functionality of a program to better self-document the block diagram window.

Double-click the icon or right-click and choose Edit Icon from the shortcut menu to display the Icon Editor shown in Figure 5.23.

This dialog box includes the following components:

Pencil—Draws and erases pixel by pixel. Press the <Shift> key to draw horizontal or vertical lines.

Line—Draws straight lines. Press the <Shift> key to draw horizontal, vertical, and diagonal lines.

Color Copy—Copies the foreground color from an element in the icon

Fill—Fills an outlined area with the foreground color

Rectangle—Draws a rectangular border in the foreground color. Double-click to frame the icon in the foreground color.

Filled Rectangle—Draws a rectangle with a foreground color frame and filled with the background color. Double-click to frame the icon in the foreground color and fill it with the background color.

Select—Selects an area of the icon to cut, copy, move, or make other changes. Double-click and press the <Delete> key to delete the entire icon.

A Text—Enters text into the icon. Double-click to select a different font. While text is active, you can move the text by pressing the arrow keys. Windows Small Fonts works well in icons.

▬ Foreground/Background—Displays the current foreground and background colors. Click each rectangle to access a color picker from which you can select new colors.

> Show Terminals—Place a checkmark in this checkbox to show an outline of the Connector Pane on the editing area. The Connector Pane is only a guide. It does not appear in the final icon.

> Copy from—Copies from a color icon to a black-and-white icon and from a black-and-white icon to a color icon.

We can drag a graphic from another file and drop it in the upper-right corner of the front panel or block diagram windows. LabVIEW converts the graphic to a 32 × 32 pixel icon. Depending on the type of monitor that might be used with the program, we can design a separate icon for monochrome, 16-color, and 256-color mode. LabVIEW uses the monochrome icon for printing unless we have a color printer.

The Icon Art Glossary on ni.com button will navigate to a web page that contains standard graphics for icons.

Customize the graphic similar to Figure 5.24, and copy it to the black-and-white and 16 color icons by clicking them and choosing copy from 256 colors. Click the OK button and save the VI.

To connect wires into the subprogram controls from and out of the subprogram indicators in the calling program, we must define the subprogram terminals in the connector pane. The connector pane defines the inputs and outputs wired to and

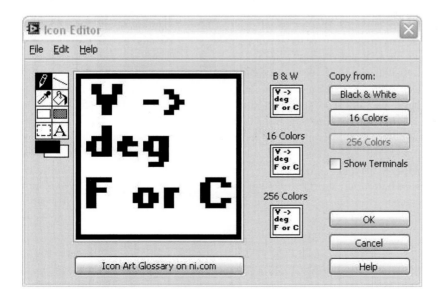

FIGURE 5.24
Temperature Units Program Custom Icon

FIGURE 5.25
Temperature Units
Conversion program
connector pane

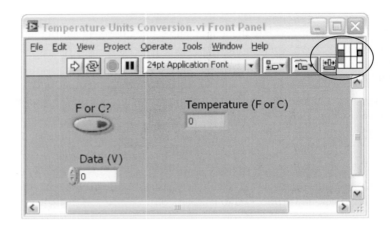

from a subprogram. It is a set of terminals that corresponds to the controls and indicators.

Display the connector pane and its connector pattern by right-clicking on the icon from the front panel window and choosing Show Connector. LabVIEW generally displays a pattern that has one terminal for each control or indicator on the front panel window. You can choose another pattern by right-clicking on the connector pane and choose Patterns. Figure 5.25 shows the connector pane for the Temperature Units Conversion VI with input terminals for Data and units and an output terminal for temperature value. The reason we added the controls and indicators above is to create these terminals. When we place the subprogram icon into another program, we will wire into and out of these terminals.

Define connections by assigning a front panel control (or indicator) to each of the connector pane terminals. Use the rectangles in the connector pane to assign inputs and outputs.

To assign the terminals in the Connector Pane:

FIGURE 5.26
Mouse change to
wiring tool for
Connector Pane

FIGURE 5.27
Connector Pane
terminal turns black

1. Move the mouse over the terminal to be assigned. The mouse will change to the wiring tool shape (Figure 5.26).
2. Left-click on the terminal. It will turn color from white to black (Figure 5.27).
3. Left-click the control (or indicator) to assign that control (or indicator) to the previously chosen terminal on the front panel window (Figure 5.28). The terminal will turn color to match the control (or indicator) data type.
4. Continue this process until all terminals have been assigned.

FIGURE 5.28 Connector pane
terminal color matches data type

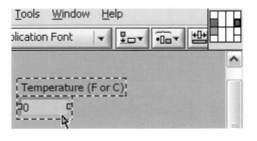

5. Right-click the connector pane and choose Show Icon.
6. Save the VI.

Add the subprogram to the Temperature Measurement State Machine, as shown in Figure 5.29:

> Choose the Select a VI from the Functions palette, as shown in Figure 5.29, and navigate to the sub VI file.
> Place the sub VI icon on the block diagram window to create a subprogram call.
> Right-click the F or C? input terminal and create a control.
> Complete the wiring to the subprogram as shown in Figure 5.29.

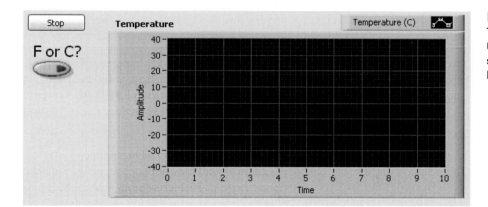

FIGURE 5.29
The Temperature Units Conversion subprogram in the Display state

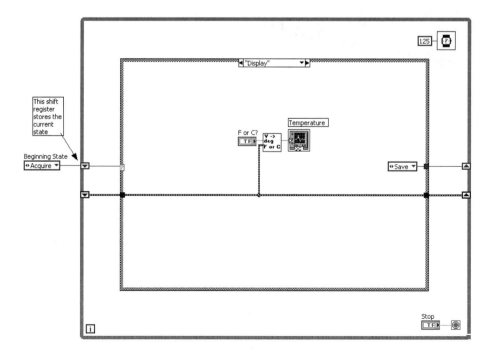

> ➤ Arrange the control on the front panel window.
> ➤ Save the Temperature Measurement State Machine and test it with F and again with C units chosen. Open the file and examine the contents.

Alternatively, if the subprogram is open, we can drag the icon from the subprogram front panel window onto the block diagram window of the calling program.

If we use a group of graphical subprograms together often, we can use a consistent connector pane with common inputs in the same location to simplify the wiring patterns. Furthermore, we can designate which inputs and outputs are required, recommended (the default), and optional to help users to remember to wire sub VI terminals by right-clicking a terminal in the connector pane, selecting "This Connection Is" from the shortcut menu, and checking Required, Recommended, or Optional. The Run arrow will be broken if you do not wire the Required inputs. In the Context Help window, the labels of Required terminals appear bold, Recommended terminals appear as plain text, and Optional terminals appear dimmed, or do not appear at all, and the wire stub is shorter. The labels of Optional terminals do not appear if you click the Hide Optional Terminals and Full Path button in the Context Help window. We used "recommended" for the inputs to the Temperature Units Conversion subprogram.

We can edit a subprogram by double-clicking the icon on the block diagram window. When we save the subprogram, the changes affect all calls to the subprogram, not just the current instance.

When LabVIEW calls a subprogram, ordinarily the subprogram runs without displaying its front panel window. If you want a single instance of the subprogram to display its front panel window when called, right-click the icon and select Sub VI Node Setup from the shortcut menu. If you want every instance of the subprogram to display its front panel window when called, select File >> VI Properties, select Window Appearance from the Category pull-down menu, and the Customize button and check Show Front Panel when Called and check Close Afterwards if Originally Closed. The front panel is not normally loaded into memory when the sub VI executes unless it is configured to show the sub VI front panel window.

ICONS OR EXPANDABLE NODES

You can display VIs and Express VIs as icons or as expandable nodes. Expandable nodes appear as icons surrounded by a colored field. Sub VIs appear with a yellow field, and Express VIs appear with a blue field. Use icons if you want to conserve space on the block diagram window. Use expandable nodes to make wiring easier and to aid in documenting block diagrams. By default, sub VIs appear as icons on the block diagram window, and Express VIs appear as expandable nodes.

To display a sub VI or an Express VI as an icon, right-click and select View As Icon from the shortcut menu. Remove the checkmark from View As Icon option to display the sub VI or Express VI as an expandable node. You cannot convert all existing sub VIs displayed as icons to expandable nodes. You cannot place Express VIs on the block diagram as icons by default. You can resize an expandable sub VI or Express VI.

SEQUENCE STRUCTURE

Sequence Structures should be used sparingly. The information is included here in case you inherit code that uses them, need them for embedded applications, or to answer questions on a National Instruments Certification Examination. They break the LabVIEW data-flow paradigm and reduce LabVIEW's ability to optimize execution. If the sequence contains a large number of frames, they slow execution because all frames must be executed before the program can continue.

A Sequence structure contains one or more subdiagrams, or frames, that execute in sequential order. There are two types: the Flat Sequence structure and the Stacked Sequence structure.

The Flat Sequence structure (Functions >> Structures Palette) displays all the frames at once and executes the frames from left to right until the last frame executes.

The Stacked Sequence structure (Functions >> Structures Palette) stacks each frame so we see only one frame at a time and executes frame 0, then frame 1, and so on until the last frame executes. The Stacked Sequence structure returns data only after the last frame executes. The sequence selector identifier is similar to a Case structure label. It appears at the top of the Stacked Sequence structure and shows the current frame number and range of frames.

The tunnels of Stacked Sequence structures can have only one data source, unlike Case structures. The output can emit from any frame, but data leave the Stacked Sequence Structure only when all frames complete execution, not when the individual frames complete execution. As with Case structures, data at input tunnels are available to all frames. To pass data from one frame to any subsequent frame of a Stacked Sequence structures, use a sequence local terminal. An outward-pointing arrow appears in the sequence local terminal of the frame that contains the data source. The terminal in subsequent frames contains an inward-pointing arrow, indicating that the terminal is a data source for that frame. You cannot use the sequence local terminal in frames that precede the first frame where you wired the sequence local.

SUMMARY

- Structured programming builds programs that are
  - Readable
  - Scalable
  - Maintainable
- The top-down stepwise refinement process develops enough detail to support implementation in graphical code.
- LabVIEW provides a tool called Execution Highlighting that will slow down the program and show us the sequence of actions.
- Execution order in LabVIEW is controlled by a concept called data flow. In data flow, an icon or function does not execute until it has data from all of its inputs.
- State machines help decompose complex problems and implement complex decision-making algorithms.
- LabVIEW state machines are based on a "Moore machine," which performs a specific action or task for each state in the diagram. Each state can lead to one or multiple states and can also end the process flow.
- State machines are scalable.

➤ State machines have compact code that fits within the monitor screen, which maintains readability for extremely large and complex VIs.

➤ State diagrams and state tables supplement pseudo code and flowcharts when planning state machine VIs.

➤ State machines are built from a While Loop, a Case structure, and shift registers.

➤ Enumerated type controls, or enums, are effective in self-documentation when using Case structures. They are numeric objects that associate numeric values with strings and display the string to the user.

➤ When used with an enumerated type and other data types, the Case structure changes from a Boolean valued structure with two subdiagrams to a multiple-selection structure that is similar to the C switch statement.

➤ State machines use a special feature associated with loops—the shift register—to transition between states, that is, to control flow from one state to another.

➤ We can wire into and out of shift registers. The data passed into the shift register is stored in computer memory, overwriting previous data. The data passed out of the shift register is read from the memory.

➤ Shift registers should be initialized.

➤ Each state contains state transition code that decides which state should be executed next. The code might be very simple and contain only a constant.

➤ Test large projects incrementally as new features are added. State machines and sub VIs facilitate incremental testing.

➤ A subprogram is any program that is embedded into (or called from the block diagram window of) another program.

➤ Subprograms create a hierarchical structure that supports good programming practice. When we begin with high-level pseudo code and refine it using the stepwise top-down approach, we are building a hierarchy that can be implemented with subprograms.

➤ Subprograms support incremental testing, because each can be tested individually before it is added to a project.

➤ Subprograms support teamwork, as team members can individually build subprograms that are assembled into a project.

➤ Subprograms support code reuse, as we can use them in other programs with minimal modifications if they are designed properly.

➤ Each LabVIEW graphical program contains an icon and connector pane. The Icon identifies the subprogram on the block diagram window of the calling program.

➤ The connector pane defines the terminals for passing data into and out of the subprogram when it is placed on the block diagram window of a calling program.

EXERCISES

1. Read the chapter carefully. Build and test the Temperature Measurement State Machine Program while you are reading. Update your notebook, including new shortcuts and data types.

2. Run the Temperature Measurement State Machine with execution highlighting. Write a description of the step-by-step program flow in a word-processing program and save the file. As part of the explanation, reverse the program flow so save is first, display second, and acquire third. Explain how you changed the flow and what results the program gave when you did it. Change the flow back to the original order.

3. Put a control on the front panel window for the time delay. Run the program with various time delays. What is an appropriate time delay for temperature? How fast does temperature data change?

4. Read the application note "Application Design Patterns: State Machine" on www.ni.com. Download and study the Coke Machine example program. Run the program and test it. Run it with Execution Highlighting. Write pseudo code and flowcharts for the program.

5. Create a State Machine from the Standard State Machine Template where a square LED indicator on the Front Panel turns on and stays on for 2 s before it turns off, then it turns off for 2 s, then on for 3 s, then off for 3 s, each in a different state, and then repeats this cycle until the user terminates the program. Display the total time elapsed by the program and the number of while loop iterations. Add software control timing that will stop the program automatically after a user-entered time.

6. Use the Standard State Machine template to build a simple calculator. Each calculation will be a separate state in the State Machine. The user will choose one of the following calculations: add, subtract, multiply, divide. The user will input two numbers A and B with a horizontal pointer slide control. Provide an enumerated control on the front panel window for the user to select the operation. Test with known values.

7. Build a subprogram that accepts a temperature input in either C, K, or F and converts the input to any of the other units. Save it as Temperature Conversion CKF.

8. Add the subprogram to the Temperature Measurement State Machine and add front panel controls to give the user the option of choosing F, C, or K temperature units.

9. Add an Analysis state to the Temperature Measurement Program with features that determine and display the highest (maximum) and lowest (minimum) temperature measured. Hint: Store the maximum in one shift register and the lowest in another and replace them when a higher or lower value is measured.

10. Add a feature to the Temperature Measurement Program Analysis state that displays the average temperature measured.

11. Build a subprogram that solves a strain gage bridge circuit and outputs ΔR, if the bridge has four gages, using the equation

$$\Delta R = (Vo * Ro)/Vs$$

where ΔR is the change in resistance due to loading, Vo is the measured value, Ro is the resistance of the unloaded strain gage, and Vs is the supply voltage. Refer to Figure 3.17 for an example bridge circuit. The user will enter a value for the supply voltage and the resistances in the main program. The main program will measure voltage out and pass it to the subprogram. Save the program as Strain Gage Bridge. vi. Do some manual calculations to get some values for testing. Run the program and compare the results with the manual calculation results.

12. Experiment with the connector pane terminal designations of recommended, required, and optional with the strain gage subprogram. Designate a terminal of each type and observe the difference between them in the Context Help window. Add the subprogram to a block diagram window. Explain what happens if the required terminal isn't wired.

13. Use the LabVIEW Example Finder to locate and open the Time to Match.vi. Save it to c:\temp. Reprogram it using the state machine architecture to replace the Sequence structure. Save it as Time to Match State Machine.vi.

14. Get a partner to exchange programs with. Modify the simple calculator program block diagram built above. Make it difficult for someone else to understand by breaking the programming style rules. Save it as calculator bad one NNN.vi, where NNN are your initials. Trade it with your partner's program. Modify the traded program to fix it, following the programming style rules, and save it as calculator good one.

Arrays

INTRODUCTION

Previous chapters explained how to design and develop DAQ applications that can produce large amounts of data. This chapter explains how to manage large data sets with array and cluster data structures. An example state machine is developed that simulates climatic data using a For Loop, a Formula Node, graphs and charts, clusters, and statistical analysis. The chapter explains caveats with LabVIEW's ability to automatically coerce data types. Up to this point, we have used Express VIs, but more versatile applications can be developed with lower-level functions for communicating with external applications. These functions are explained in Chapter 7.

OUTLINE

ARRAYS

One of the major advantages of using computers is that they can process large amounts (millions of data points) of data quickly. We apply this ability to measuring and controlling the physical environment around us, simulating measurements and physical phenomena, building multidimensional representations like topographic maps, and so on. When writing programs that access and store large volumes of data, we need structures that can access individual elements of data within the set as well as the entire set. The term *data structure* refers to a scheme for organizing related pieces of information. There are several types of data structures. We will use two structures to organize data in this chapter: arrays and clusters. These structures store data such as coefficients of linear equations, vectors, matrices, coordinates, data bases, and so forth; consequently, they are very popular and useful.

Up to this point in the text, we have given each control, indicator, and constant a separate variable or constant identifier or label so we could access their locations in computer memory. If we store millions of data points in computer memory, we don't want to give each of them a different label. It is more efficient to assign a label to a block of memory containing the data and associate all the individual values with the block. Such a data structure is called an array. LabVIEW arrays can have as many as $2^{31} - 1$ elements per dimension, memory permitting.

A one-dimensional array can be visualized as an ordered list, such as a row or column of numbers. A two-dimensional array can be visualized as a matrix or a spreadsheet. A three-dimensional array example is a topographic map with a value of height at each coordinate pair, where the coordinate pairs form a two-dimensional array. We can build arrays of numeric, Boolean, path, string, waveform, and cluster (to be defined later in this chapter) data types, but all elements in the array must be the same data type in LabVIEW.

An array consists of an identifier, elements, dimensions, and indexes. The identifier is the label. The identifier is used to access the entire block of data. Elements are the individual values of data in the array. An index is used to locate the position of elements in the array. Indexes are equivalent to subscripts or offsets. Dimensions refer to the length, height, or depth of an array. For example, we might have millions of temperature measurements in °C that we wish to place in a one-dimensional array. We will give the array the identifier "temperature data." Then we can access individual values with the identifier and index (in parentheses), as shown for the first five values of the array in Table 6.1. Arrays are zero indexed; that is, the index of the first data value is 0, not 1.

DEFINING AN ARRAY ON THE FRONT PANEL

We must communicate with the computer operating system to reserve space in memory for data structures. We do this by placing controls or indicators on the front panel window of a program or programmatically defining the structures on the block diagram window. Creating an array from the front panel window is two-step process.

TABLE 6.1 **One-Dimensional Array of Temperature Values**

Identifier and Index	Value
Temperature data (0)	25
Temperature data (1)	28
Temperature data (2)	22
Temperature data (3)	26
Temperature data (4)	30

First, place an array shell (Controls >> Array, Matrix & Cluster) on the front panel window (Figure 6.1), and second, insert a control or indicator of the desired data type into the shell (Figure 6.2). In this case, we created an array of double-precision floating-point elements by dragging in a numeric control into the array shell, but we could have used Boolean, string, path, refnum (discussed in Chapter 7), or cluster (discussed later in this chapter) types of controls or indicators. Note that the Array shell automatically resizes to accommodate the new object. If you are coding in Lab-VIEW while studying this text, you will notice that initially the shell icon was black on the Block Diagram (signifying the array data type was undefined), but after you inserted the Numeric Control, it defined the data type (double-precision floating point), and the block diagram icon color changed to orange.

FIGURE 6.1 Array shell

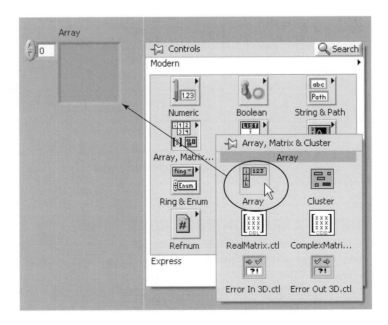

Index Display

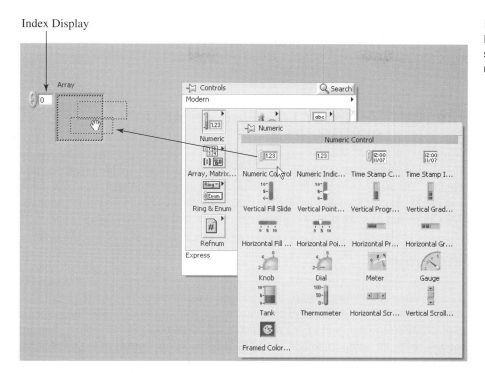

FIGURE 6.2
Defining an array
shell data type with a
numeric control

We can use the control to enter values into an array. Carefully position the cursor and drag the lower edge of the array control down to display all five elements, as shown in Figure 6.3, and enter the values 25, 28, 22, 26, and 30. The index display shown on the left side of the front-panel Numeric-array control in Figure 6.2, allows us to access any element. To use the index control don't drag the lower edge to display all five elements; just display the first element. Enter 25 in the first element (index 0), change the index to 1 and enter the next value, and so on, until all five values are entered.

To create a multidimensional array, right-click the index display on the front panel window and select Add Dimension from the shortcut menu.

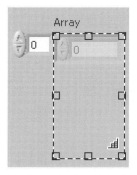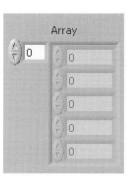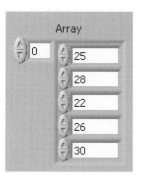

FIGURE 6.3
Assigning values to
an array control

Indexes let us navigate through an array and retrieve elements, rows, columns, and pages. A 2-D array, for instance, would be displayed as several rows or columns, and each element would have a row index and a column index. The indexes are listed in (row, column) order.

PROGRAMMATICALLY CREATING AN ARRAY ON THE BLOCK DIAGRAM

The most common way to create an array is programmatically on the block diagram window. The array could result from measurement data or simulated data. Simulated data can be useful if we are developing a program on a computer that does not have DAQ hardware or if we want to compare simulated and measured data.

FOR LOOP

We used While Loops for repetition in the programs we developed previously. Figure 6.4 introduces the For Loop repetition structure, which has a different border, but uses the same iteration terminal. Add a For Loop (Functions >> Structures), to the block diagram window using the same technique as adding a While Loop. Instead of the conditional terminal, the For Loop has a loop-count terminal designated N in the upper-left corner. The For Loop repeats N times and exits. The VI in Figure 6.4 builds an array of 10 random numbers using the Random Number function (Functions >> Numeric >> Random Number). If you are coding in LabVIEW while reading this text, you will notice that N is blue indicating integer data type.

BUILDING ARRAYS WITH LOOP TUNNELS

To create an array of the 10 random numbers, wire from the output of the Formula Node to the For Loop border to create a tunnel through the border, as shown in Figure 6.4. The loop tunnel represents computer memory. The value(s) at the loop tunnel are available in computer memory to subsequent icons on the block diagram window. The memory can store either the values from all iterations of the loop in an array, or it can store only the value from the last iteration. To write all of the loop

FIGURE 6.4
Building an array of
random numbers
programmatically

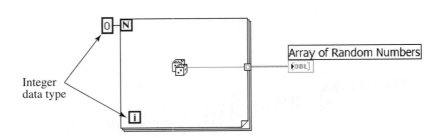

Integer
data type

values to an array, enable indexing, which is the default For-loop tunnel condition. To write only the last value, right-click on the tunnel and disable indexing. Loop tunnels are available on While Loops also, but by default, the While-loop tunnel has indexing disabled.

FORMULA NODE

Sometimes we want to simulate data in addition to, or instead of, measuring a phenomenon. For example, we might want to simulate temperatures near Denver, Colorado, where the normal temperature range is 11.7°C (53°F) to −3.9°C (25°F) in November. We might use this information to design a measurement system, compare with our measurements, replace missing data, or make a future projection. Let's assume the temperature variation follows a sinusoid pattern. We would like to create an array that can be graphed, as shown in Figure 6.7. The range is 11.7 − (−3.9) or 15.6°C. The amplitude of the sinusoid will be one-half of the range, or 7.8°C. Consequently, the sinusoid will be centered about 3.9°C (11.7°C − 7.8°C).We will generate the 24 simulated, hourly data points with a sinusoid with the peak at the time of maximum temperature. There are many ways to generate sinusoid signals in LabVIEW:

Simulate Signal Express VI

Sine Function

Sine Wave and Sine Pattern VIs

Sine WavePtByPt VI

Sine Waveform

Sine equation in a Formula Node

We will simulate a sinusoid with an equation in a Formula Node. Even though our problem is not very complicated, we will use this opportunity to learn about the Formula Node. Begin by placing a Formula Node on the block diagram window (Functions >> Structures). The Formula Node is a resizable box similar to the For Loop, While Loop, and Case Structure. The Formula Node contains one or more C-like statements delimited by semicolons, as shown in Figure 6.5. As with C, add comments by enclosing them inside a slash/asterisk pair (/*comment*/).

We can develop an equation or a set of equations in the Formula Node to simulate hourly temperature data points over one day. We will use the equations:

$$t = (i/24) * 2 * pi;$$
$$T = A * \sin(t) + O;$$

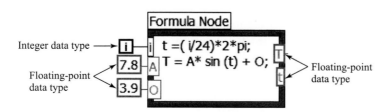

FIGURE 6.5
Formula Node

where *t* is the time converted to an angle in radians (the sine function requires values in radians).

i is the For Loop iteration count, which will vary from 0 to 23

T is the temperature (°C) at each of the 24-hour samples.

A is the sine amplitude (°C) that is one-half the thermal range of 15.6°C (11.7°C − (−3.9°C)), or 7.8°C.

O is the offset from 0 of 3.9°C (11.7°C − 7.8°C).

First, we will declare input variables by right-clicking the left side of the Formula Node border and selecting Add Input from the shortcut menu. We need to add inputs for *i*, *A*, and *O*. Connect the iteration count terminal to *i*. Right-click on *A* and *O* and create constants with the appropriate values shown in Figure 6.5. If you are programming in LabVIEW while reading this text, you will notice that the *i*nput turns blue to indicate integer data but the *A* and *O* inputs are orange floating-point data.

Next, type the equations into the Formula Node, ending each statement with a semicolon, as shown in Figure 6.5. It is important to use LabVIEW Help when working with the Formula Node. Choose Search the LabVIEW Help from the Help pull-down menu. Choose the Index tab and type "Formula Node" to see a list of topics. Topics such as functions and descriptors, operators, and syntax provide essential information.

Finish the Formula Node by declaring the variables *t* and *T*. To define an output variable, right-click the right side of the Formula Node border and select Add Output from the shortcut menu. The output variable name must match either an input variable name or the name of a variable declared inside the Formula Node. We do not have to wire all output variables, but we must wire all input variables. We must define all variables even when we do not wire to them. We could have used the Express Formula VI (Functions >> Express >> Arithmetic & Comparison) instead of the Formula Node. It uses a calculator interface to create mathematical formulas similar to those performed by most basic scientific calculators. Refer to the Calculation on Dynamic Data Example program (Help >> Find Examples).

Expression Nodes can be used in place of Formula Nodes to calculate expressions, or equations that contain a single variable. Expression Nodes use the value you pass to the input terminal as the value of the variable. The output terminal returns the value of the calculation. Refer to Help for more information on Expression Nodes.

AUTO-INDEXING

If you wire an array to a For Loop or While Loop input tunnel, you can read and process every element in that array by enabling auto-indexing. When you wire an array to an input tunnel on the loop border and enable auto-indexing on the input tunnel, elements of that array enter the loop one at a time, starting with the first element. When auto-indexing is disabled, the entire array is passed into the loop. If you enable auto-indexing on an array wired to a For Loop input terminal, LabVIEW sets the count terminal to the array size so you do not need to wire the count terminal. Because you can use For Loops to process arrays an element at a time, LabVIEW enables auto-indexing by default for every array you wire to a For Loop.

This is a common feature when working with arrays and the Formula Node. Rather than wire an input array to an Input to the Formula Node, put a For Loop around the Formula Node and auto-index the array to process the elements one at a time. Then pass the elements through an indexing-enabled tunnel out of the For Loop to rebuild the array.

GRAPHS AND CHARTS

To display the result, right-click on the array and create an indicator that automatically appears on the front panel window. Switch to the front panel window and expand the indicator to display the entire array, as shown in Figure 6.7. To graph the array, add a Waveform Graph (Controls >> Graph >> Waveform Graph) to the front panel window. Wire to it on the block diagram window, as shown in Figure 6.6. Save the program as Temperature Array. Run it to populate the array indicator and the graph with simulated temperature values.

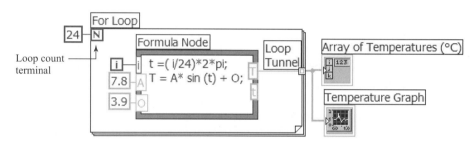

FIGURE 6.6
Temperature array VI block diagram window

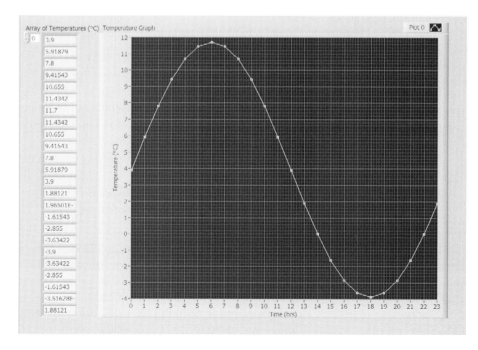

FIGURE 6.7
Graph of temperature array

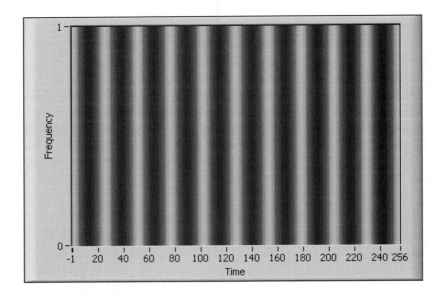

The Graph palette contains both graphs and charts. Graphs and charts differ in the way they display and update data. Programs with graphs usually collect the data in an array and then plot the data to the graph, which is similar to a spreadsheet that first stores the data, then generates a plot of it. In contrast, a chart appends new data points to those already in the display to create a history. The history data remains in memory after we run a VI. We can clear the history by right-clicking on the chart and choosing Data Operations >> Clear Chart.

Other types of graphs and charts in LabVIEW include the following:

XY Graphs plot *x,y* points where *x* can be time or some other variable. For example, XY Graphs can plot torque versus speed or stress versus strain.

Intensity Graphs or Charts plot 3-D data on a 2-D plot, as shown in Figure 6.8. For example, an intensity chart could be used to plot temperature distribution over a surface. If you open the Intensity Chart example VI in LabVIEW, you will notice the third dimension value is displayed in shades of color.

Mixed Signal Graphs plot waveform data, *x-y* data, digital data, and clusters of data, as shown in Figure 6.9.

3-D Graphs show data in 3D with three axes, as shown in Figure 6.10. If you open the 3D graph Example VIs in LabVIEW you will notice the plot surfaces are shaded colors.

Examples of the various graphs and charts are available in LabVIEW Example Finder, found under the Help >> Find Examples menu. There, select Browse By Task and choose Fundamentals >> Graphs and Charts. Instructions on using the various graphs and charts are available in LabVIEW Help > Graphs and Charts.

Graphs and charts are extremely valuable to program users. Consequently, LabVIEW provides several options for customizing them to the user's needs.

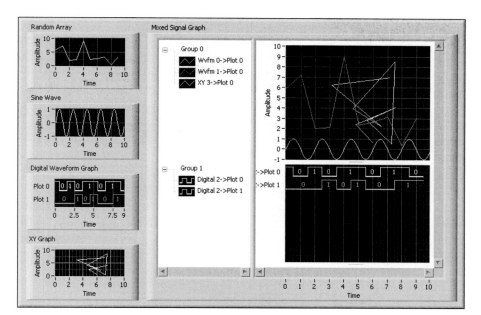

FIGURE 6.9
Mixed Signal Graph example (LabVIEW Help > Types of Graphs and Charts)

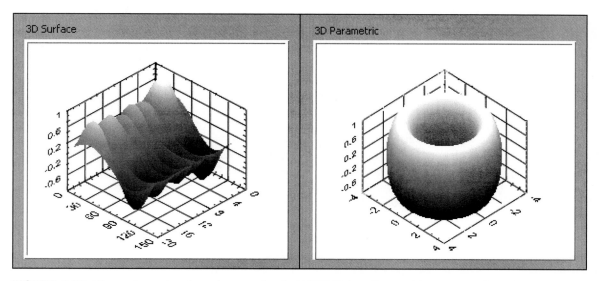

FIGURE 6.10 Three-dimensional graph examples (LabVIEW Help > Types of Graphs and Charts)

Right-click the graph or chart and select Visible Items from the shortcut menu to display or hide the following options:

➤ Plot Legend—Defines the color and style of the plot(s). Resize the legend to display multiple plots.

➤ Scale Legend—Defines labels for scales and configures scale properties.

> ➤ Graph Palette—Changes scaling and formatting while a VI is running.
> ➤ X Scale and Y Scale—Formats the x- and y-scales. By default, Auto-scaling is enabled for graphs. Turn Auto-scaling on or off by right-clicking the graph and selecting X Scale > Auto-scale X or Y Scale > Auto-scale Y from the shortcut menu. Auto-scaling is off in Figure 6.7. The *x*- and *y*-axes ranges were edited to reflect the maximum and minimum temperature values and 24 hour cycle.
> ➤ Cursor Legend (graph only)—Displays a marker at a defined point coordinate. You can display multiple cursors on a graph.
> ➤ X Scrollbar—Scrolls through the data in the graph or chart. Use the scrollbar to view data that the graph or chart does not currently display.

In addition to making the preceding list of items visible, we can further customize graphs and charts with the properties dialog box. Open it by right-clicking on the graph or chart and choose Properties from the shortcut menu. Some common properties we can change include the following:

The format, precision, and mapping mode.

Grids, grid lines, and grid colors.

X-axis and y-axis appearance. For example, the x-axis default configuration uses floating-point notation with a label of Time, and the y-axis is configured to use automatic formatting with a label of Amplitude.

Charts can display in different modes, such as strip chart, sweep chart, and scope chart. Right-click the chart and select Advanced >> Update Mode to change modes.

SPECIAL CHARACTERS

We may need to use special characters such as a trademark (™) or degree (°) symbol in a program: for example, the degree (°) symbol in the Figure 6.7 y-axis title. Special characters are available in the Windows Character Map. To access it, click **Start** on the Windows Desktop, point to **Programs**, point to **Accessories**, point to **System Tools**, and then click **Character Map**. The Character Map gives the keystrokes to create the character and has a character copy facility.

ARRAY FUNCTIONS

The Temperature Array program graphs the array after it is built, using the waveform graph. This is adequate for a quick simulation or for small data acquisition programs. However, we might want to display each data point as it is added to the array if we have a larger or slower application. This modification will give us the opportunity to explore some of the many functions available in LabVIEW to manipulate arrays.

Modify the block diagram window of the Temperature Array VI to build the block diagram shown in Figure 6.11. Save the VI with a new name, Temperature Array PtByPt. Do not change the front panel window.

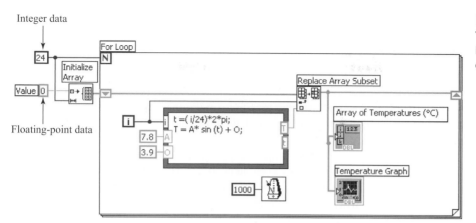

FIGURE 6.11
Temperature Array
PtByPt VI block
diagram window

Initializing or defining an array reserves memory space for the array and assigns a data type to that array. The Initialize Array Function creates an n-dimensional array in which every element is initialized to the value wired to the element input terminal. To initialize the Temperature Array, add the Initialize Array Function (Functions >> Array) and create a constant for the dimension size input. Enter 24 for the constant, because we need a one-dimensional array with 24 elements for our temperature values. Create another constant for the element input and enter 0 to place a 0 value in each of the 24 elements. If you are coding in LabVIEW while studying this text, you will notice that the dimension size input is blue representing integer data type and the element input is orange representing floating-point data type.

To place the initial Temperature Array into memory that is accessible from the For Loop, create a shift register on the For Loop border and wire the array output from the Initialize Array function into the shift register.

To replace the 0 element values with simulated temperatures, add the Replace Array Subset function (Functions >> Array), as shown in Figure 6.11. This function replaces an element or sub-array at the point specified by the index input. Wire the array from the shift register to the array input terminal, as shown in Figure 6.11. Wire the T output from the Formula Node to the new element terminal of the Replace Array Subset function. Next, wire the For Loop iteration value into the index terminal. The result of this operation is that the index will increment each time the For Loop iterates. As the index increments, the initial 0 value at that index will be replaced by the value calculated by the Formula-Node equations. If we do not wire an index for a dimension, the function replaces all the elements in that dimension.

To slow down the graph update, add a Wait Until Next ms Multiple function with a constant value of 1000. Run the VI and observe that initially the graph shows 0's in every element as a result of executing the Initialize Array Function. The VI replaces the 0's as the For Loop iterates and executes the Replace Array Subset function. Run the VI once with auto-scaling on, and again with it off, to observe the effect of auto-scaling on the graph.

The preceeding program could have been written using the Build Array function instead of initializing an array and replacing elements in the array. If possible, do not programmatically create arrays using the Build Array function within a loop because the function makes repetitive calls to the LabVIEW memory manager, consuming processor resources and affecting loop timing.

Engineers and scientists commonly write programs that manipulate arrays, so LabVIEW provides many functions to support array programming:

> Array Max & Min returns the maximum and minimum values in an array.
> Array Size returns the number of elements in each dimension of an array.
> Array Subset returns an element or a sub-array of an array.
> Build Array concatenates multiple arrays or appends elements to an array.
> Decimate 1D Array divides the elements of a 1-D array into output arrays, placing elements in the outputs successively.
> Delete From Array deletes an element or sub-array from an array.
> Index Array returns an element or sub-array of an array.
> Initialize Array creates an array.
> Insert Into Array inserts an element or sub-array into an array.
> Interleave 1D Arrays interleaves corresponding elements from the input 1-D arrays into a single 1-D array.
> Interpolate 1D Array linearly interpolates a decimal y value from an array of numbers using a fractional index or x value.
> Replace Array Subset replaces an element or sub-array of an array.
> Reshape Array changes the dimensions of an array.
> Reverse 1D Array reverses the order of the elements in an array.
> Rotate 1D Array rotates the elements in an array a specified number of indexes.
> Search 1D Array searches for an element in a 1-D array.
> Sort 1D Array sorts the elements in a 1-D array in ascending order.
> Split 1D Array splits an array at a specified index and returns the two portions.
> Threshold 1D Array compares an input value, y, to the values in an array of numbers and returns the fractional index where the numbers become greater than y.
> Transpose 2D Array rearranges the elements such that a 2-D array $[i, j]$ becomes array $[j, i]$. Rows of elements become columns, and columns become rows.

Examples of the various array functions are available in Help >> Find Examples. Select Browse By Task and choose Fundamentals >> Arrays and Clusters. Instructions on using the various array functions are available in Help.

MULTIDIMENSIONAL ARRAYS

We usually acquire data from more than one transducer, and we combine several 1-D arrays into a single multidimensional array. For example, we might combine a 1-D humidity array with the 1-D temperature array to build a 2-D array called Climate Data. This associates both arrays in computer memory so they can be accessed with the Climate Data label. Individual data elements can be accessed by using their index (row, column) within the 2-D array. Each row represents a point in time where we have a temperature

TABLE 6.2   **Two-Dimensional Array of Simulated Climate Data**

Row	Column 0	Column 1
0	3.9	68
1	5.9	62
2	7.8	55
3	9.4	49

measurement and a humidity measurement. For example, in Table 6.2, row 0 contains a temperature value in °C in column 0 and a humidity value in % in column 1. The row, column index of the 7.8°C value is [2,0]. The value at index [1,1] is 62%.

We create a multidimensional indicator by expanding the array indicator to display two columns, one for temperature and the other for humidity. Right-click on the Array indicator index and choose Add Dimension to display both the row and column indexes on the array indicator. The upper index is the row index, and the lower one is the column index. Then, use the cursor to drag the upper-right corner of the array border one cell to the right to add the second column to display the humidity values, as shown in Figure 6.12.

In addition to displaying the humidity values in the array indicator, LabVIEW graphs and charts can plot each of the columns or rows of multidimensional arrays as separate lines, or plots. To add a plot of humidity data, drag the plot legend upward one cell to add a legend for an additional plot line, as shown in Figure 6.13. If the plot legend is not visible, right-click the graph or chart and select Visible Items >> Plot Legend from the shortcut menu. The plots can have different styles and colors.

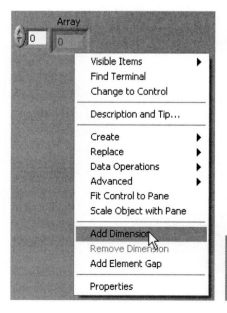

FIGURE 6.12
Creating a two-dimensional array control

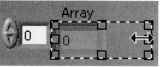

FIGURE 6.13 Drag the plot legend upward to add plots

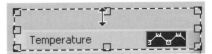

Next, add a new y scale for the new plot as the humidity values will not fit in our −4 to 12°C temperature scale. All graphs support multiple x- and y-scales, and all charts support multiple y-scales. Right-click the y-scale of the waveform graph and select Duplicate Scale from the shortcut menu, as shown in Figure 6.14. If you want to move the new scale to the opposite side of the display, right-click the new scale and select Swap Sides.

Now that we have two scales, we need to inform the computer what scale to use for each plot. This is called associating a plot with a scale. Associate each scale by right-clicking the plot in the plot legend. Choose Y Scale from the shortcut menu, then choose either the temperature or humidity scale, as shown in Figure 6.15.

FIGURE 6.14
Adding a duplicate scale

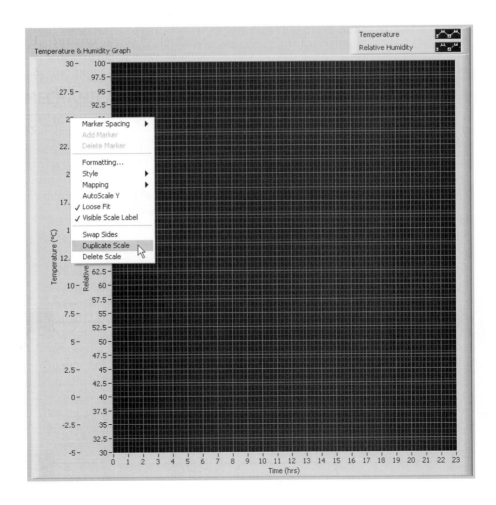

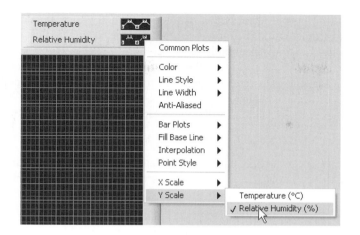

FIGURE 6.15
Associating a plot
and a scale

To change the range of the new scale, use the Operating tool or the Labeling tool to highlight the end value(s) of the scale and enter a new value. Disable auto-scaling before setting the scale manually. Adjust the range and the increments so grid lines are used by both scales, as shown in Figure 6.17, for example.

We need to expand the initialized array on the block diagram window to 2-D, also. Right-click the Initialize Array function and choose Add Dimension or drag the border to resize, as shown in Figure 6.16. Wire the 24 constant to the upper dimension-size terminal to create 24 rows, and wire the 2 constant to the lower dimension-size terminal to create two columns. If you are coding in LabVIEW while studying this text, you will notice the constants for number of rows and number of columns are blue, or integer data, and the value constant is orange representing floating point data.

Next we will generate the humidity values. Assume that relative humidity varies between 42 and 96% and its sinusoid is offset by 12 hours, or π radians, from

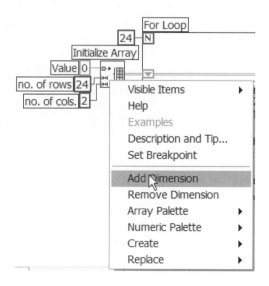

FIGURE 6.16
Resizing the Initialize
Array function

the temperature sinusoid—that is, the maximum humidity corresponded in time to the minimum temperature. Therefore, we will add a phase angle offset of π radians to the humidity equation. The amplitude is $(96\% - 42\%)/2 = 27\%$. The offset from 0 amplitude is $96\% - 27\% = 69\%$. We will modify the names of the constants to identify which are for temperature by adding T to the temperature constant name and H to the humidity constant name.

Consequently, the equation for T will become

$$T = AT * \sin(t) + OT;$$

where t is the time converted to an angle in radians $t = (i/24) * 2 * pi$ with no change from that described above. AT is the temperature sinusoid amplitude, and OT is the offset from 0.

The values for the humidity sinusoid will be generated by

$$H = AH * \sin(t) + OH;$$

where AH is the humidity sinusoid amplitude and OH is the offset from 0.

The array we built has two columns, one representing temperature, the other humidity. Each row of the 2-D array corresponds to the time that the temperature and humidity measurements occurred.

We would like to plot both temperature and humidity on the waveform graph with two separate plot lines. A multiplot waveform graph accepts a 2-D array of values, where each row of the array is a single plot (Figure 6.17). However, we generated temperature in column 0 and humidity values in column 1. Therefore, we

FIGURE 6.17
Temperature and Humidity Array PtByPt VI front panel window

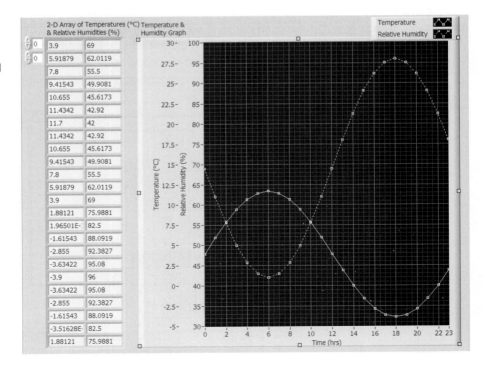

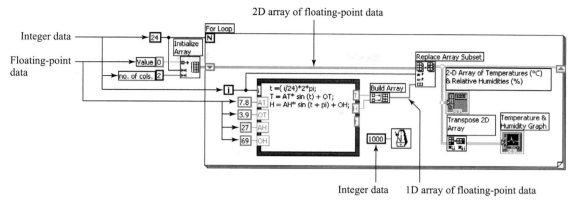

FIGURE 6.18 Temperature and Humidity Array PtByPt VI block diagram window

will use the Transpose 2-D Array function to put temperature values in row 0 and humidity values in row 1 of a 2-D array, as shown in Figure 6.18. We could also wire the 2-D array to the graph, right-click the graph, and select Transpose Array from the shortcut menu instead of adding the Transpose 2-D array function. The graph interprets the data as points on the graph and increments the x index by one, starting at $x = 0$.

COERCION

LabVIEW allows mixing data types in many functions. For example, it will add an array of double-precision floating-point values, to an array of integers. The function morphs by coercing one of the data types. A small dot will be added automatically to the terminal with the value that was coerced. This type of coercion uses extra memory as LabVIEW makes a copy of the coerced data in memory. Therefore, try to avoid it. This is more troublesome when working with large arrays as large blocks of memory can be duplicated. Additional information is available in Help >> Index >> Performance >> Memory.

STATISTICAL ANALYSIS

Most physical measurements contain noise and we can simulate it in LabVIEW. Figure 6.19 and Figure 6.20 show the results of adding individual points of uniform white noise (Functions >> Signal Processing >> Point By Point >> Signal Generation PtByPt >> Uniform White Noise) to the humidity and temperature values.

Noise is usually simulated with random numbers. Random numbers might be obtained by chance; by throwing a die, for example. If truly random numbers are generated in a sequence, they shouldn't have anything to do with each other, and they should have a specified probability of falling within a specified range of values.

Figure 6.19 depicts a noise range of -2 to 2 for temperature (°C) and -5 to 5 for relative humidity (%), with a uniform distribution, which means that each of the

FIGURE 6.19
Temperature and
Humidity Array
PtByPt with Noise
Program block
diagram window

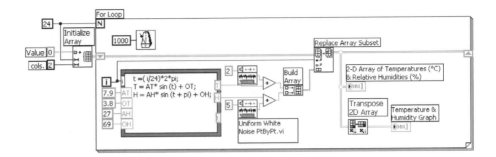

FIGURE 6.20
Temperature and
Humidity Array
PtByPt with Noise
Program front panel
window

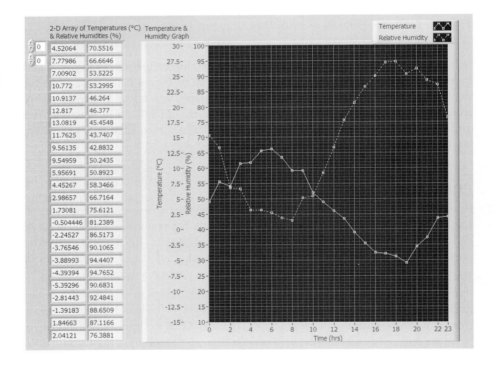

possible numbers is equally probable. The Uniform White Noise VI generates the pseudorandom sequence using a modified version of the Very-Long-Cycle random number generator algorithm. Refer to Help for additional information. Other single-point random number and noise generators are available in LabVIEW—for example, the Random number function, the Gaussian White Noise PyByPt, and Periodic Random Noise PtByPt VIs. In addition to point-by-point values, LabVIEW can generate a variety of noise waveforms as can be seen by searching the Functions palette for noise.

Engineers and scientists frequently compare simulated or actual measurement data using statistics. For example, we might want to know the mean or average value of temperature or humidity for different days during the year. The mean is the sum of the measurements y_i divided by the number of measurements (N):

$$\mu = \frac{1}{N}\left(\sum_{i=1}^{N} y_i\right)$$

We might also be interested in the middle temperature value, or median. If so, we sort the array and determine the number of values in the array, N. If N is odd, the median value occurs at index, which is the center of the stack of sorted values. If N is even, the median is the average of the two values on either side of the center of the stack.

We might also be interested in how much the values deviate from the mean value, which we measure with the standard deviation or variance. We calculate variance, σ, with

$$\sigma^2 = \frac{1}{N-1}\left(\sum_{i=1}^{N}(y_i - \mu)^2\right)$$

where $y_i - \mu$ is the difference between each measurement and the mean of all the measurements. We sum these differences, square them to assure a positive value, and divide the sum by $N - 1$. The standard deviation is the square root of the variance.

We could implement these equations in a formula node, sum values in shift registers, sort the arrays, and so on, to develop statistics of our simulated temperature and humidity arrays. However, it is easier to use VIs provided in LabVIEW, as shown in Figure 6.21 and Figure 6.22, which calculate and display statistics for temperature on one of the pages of a tab control.

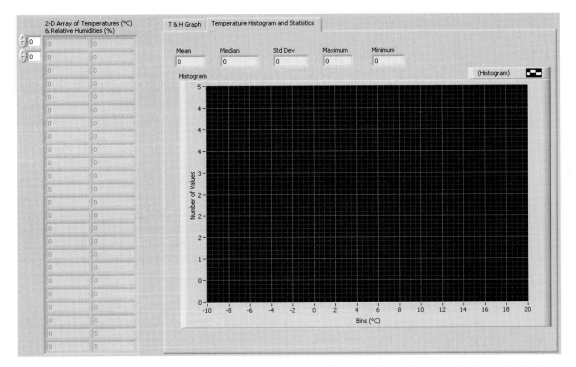

FIGURE 6.21 Temperature and Humidity Array PtByPt with Noise & Stats Program front panel window

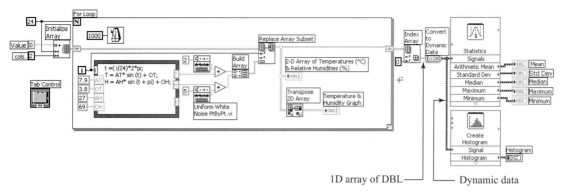

FIGURE 6.22 Temperature and Humidity Array PtByPt with Noise & Stats Program block diagram window

FIGURE 6.23
Statistics Express VI
configuration dialog
box

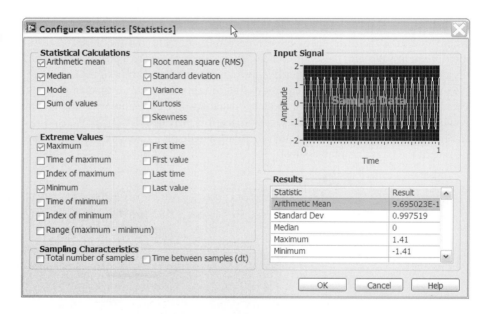

Build the block diagram in Figure 6.22, which calculates statistics for the temperature array values. To isolate temperature data for analysis, use the Index Array function (Functions >> Array) to separate the temperature values in column 0, from the 2-D array.

Next, add the Statistics Express VI (Functions >> Mathematics >> Probability and Statistics), configuring the dialog as shown in Figure 6.23. Wire the sub-array output of the Index Array function to the Statistics Express VI. The Convert to Dynamic Data function automatically appears to change the 1-D array data type to the dynamic data type required by the Express VI.

Right-click the Statistics Express VI dynamic data type output terminals and select Create >> Numeric Indicator from the shortcut menu to display the statistics data on the front panel window. Note that the dynamic data automatically changes to double-precision floating point in the indicators. This is a quick way to add front panel objects while building the block diagram. We will continue to build the block diagram window before returning to the front panel window to arrange these objects.

HISTOGRAM

LabVIEW has additional statistics VIs in the Functions >> Mathematics >> Probability and Statistics subpalette. One calculates and displays a histogram. To display a histogram of the temperature values, add the Create Histogram Express VI as shown in Figure 6.22, and configure it as shown in Figure 6.24. Right-click on the Create Histogram output terminal and select Create >> Graph Indicator from the shortcut menu to display the data in a graph on the Front Panel.

The histogram is a frequency count of the number of times that a specified interval occurs in the input sequence. The user can specify the width of the frequency bin by entering the number of bins, m. The program then determines the width, Δx:

$$\Delta x = (\text{max} - \text{min})/m$$

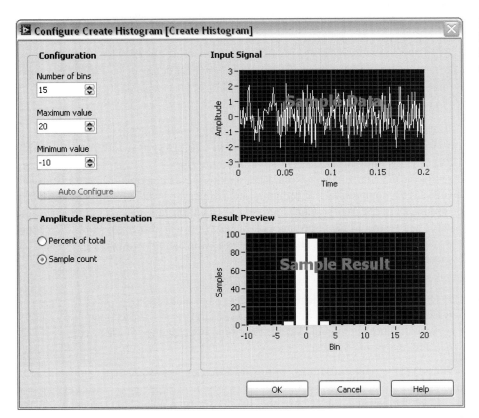

FIGURE 6.24
Create Histogram
Express VI
configuration dialog

FIGURE 6.25

Histogram example

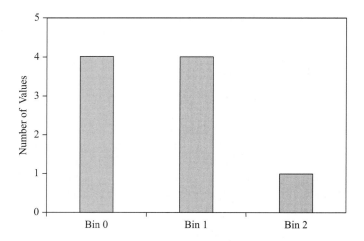

The VI sets the centers of each bin according to the following equation:

$$\text{center}[i] = \min + \Delta x/2 + i * \Delta x$$

The program calculates the histogram by summing the number of values within each bin. For example, if the input sequence is $Y = [0,1,3,3,4,4,5,5,9]$ and $m = 3$, $\Delta x = (9-0)/3 = 3$. The centers of the bins are

$$\text{Center (0)} = 0 + 3/2 + 0 * 3 = 1.5$$
$$\text{Center (1)} = 0 + 3/2 + 1 * 3 = 4.5$$
$$\text{Center (2)} = 0 + 3/2 + 2 * 3 = 7.5$$

The bin widths are Bin (0): $0-3$; Bin (1): $3.001-6$; and Bin (2): $6.001-9$.

The number of values in each bin is Bin 0: 4 [0,1,3,3]; Bin 1: 4 [4,4,5,5]; Bin 2: 1 [9]. The resulting histogram is shown in Figure 6.25.

Place a tab control on the front panel window so we can display the statistics and histogram without deleting or reducing the Temperature and Humidity values array or graph. Drag the Temperature and Humidity Graph into the tab control, and the control will resize. Relabel the pages of the tab control as shown in Figure 6.21. Drag the statistics indicators and Histogram graph onto the appropriate page of the tab control as shown.

Customize the graph:

> Right-click on the plot legend and select a bar chart style.
> Set the y-axis precision in graph properties to 0 digits of precision.
> Turn off auto-scaling and set the lower and upper values of the y-axis.
> Label the x- and y-axes.

LabVIEW Help contains more information on statistical analysis with LabVIEW.

Save the VI and run it. Are the statistics values what you expect from the Temperature and Humidity Array values?

CLIMATE SIMULATION STATE MACHINE

Our program is more readable, maintainable, and scalable if we convert it to a state machine. We will divide the application into the following states:

Generate

Graph

Analyze

The application will execute these four sequentially 24 times to simulate hourly temperature and humidity measurements for one day and stop. Figure 6.26 shows the state diagram for the application. State names and the number of states vary with user preference. For example, we could add initialize and stop states and still have a perfectly correct state machine.

We can modify the existing program or copy the code into the Standard State Machine Template. In either case, the front panel window will not change as shown in Figure 6.27.

We will code the Generate case, as shown in Figure 6.28. If we use the Standard State Machine Template,

1. Delete the free labels from the template.
2. Remove the wire, stop constant, and equal operator connected to the While Loop conditional terminal.
3. Then, copy the appropriate code from the Climate Simulation program and place it as shown.

The program generates simulated temperature and humidity values in this state that must be shared with other states for graphing and analyzing. Therefore, wire the array of values into a shift register.

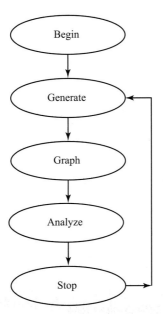

FIGURE 6.26
Climate Simulation state machine state diagram

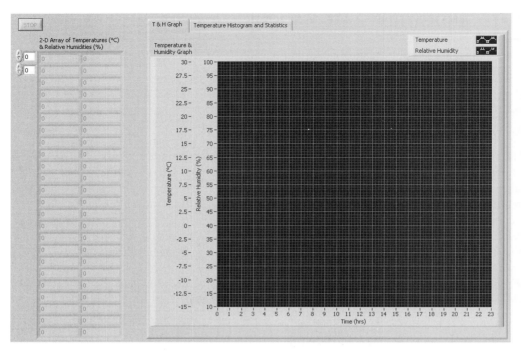

FIGURE 6.27 Climate Simulation State Machine Front Panel

FIGURE 6.28
Generate case

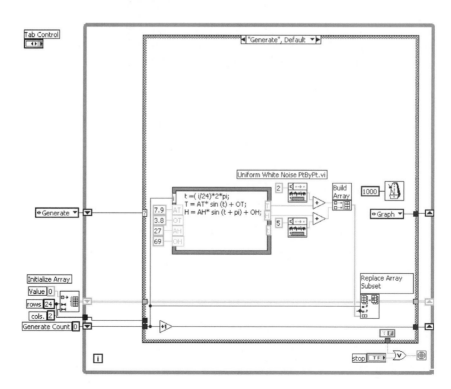

The While Loop will iterate each time we change cases, so the value of i is no longer useful for the array index or for stopping the program after 24 executions of the generate state. Therefore, add a generate count constant that increments by one each time the Generate state replaces a row in the array. Initialize the value at 0 and wire it to a shift register as shown in Figure 6.28 so all cases can access it.

Wiring output tunnels in Case Structure subdiagrams can create messy code. Therefore, move the shift registers to the bottom of the case structure so their wires won't run through the code. There are three scenarios for creating these wires:

1. Writing to memory. The values in the temperature and humidity 2-D wire change in the Generate case. So the code that produces the new value sends it via a new wire to the Case Structure output tunnel and overwrites the shift-register memory.
2. Reading from memory. Because the values don't change, we wire from the input tunnel to the output tunnel. Then, we wire a branch to the VI or function that will use the data. We read what is in the shift-register memory, but we don't change the memory.
3. Neither reading nor writing. The values don't change, so we wire from the input to the output tunnel as in the preceding. For example, examine the generate count wire in the Graph case. This practice satisfies the rule of wiring to all of the Case Structure output tunnels.

The program should execute the Graph state after executing the Generate state. So, copy and paste the Climate States enumerated constant, change its value to Graph, and wire it to the right state-transition shift register.

We want the program to stop after processing 24 rows of data or if the user presses the Stop button. Add an Or function in front of the, While Loop conditional terminal as shown in Figure 6.28. Create a Boolean constant with a False value in the Generate case and wire through a tunnel to the Or function. Create a Stop control for the front panel window. Wire it to the other terminal of the Or function as shown. We will add code to the analyze State to stop after 24 rows.

After coding the Generate State, we will code the Graph State. Increment the case selector label to reveal the Graph State subdiagram. Copy code from the Climate Simulation program and place it in the subdiagram, as shown in Figure 6.29.

To read and display the data, wire from the 2-D array shift register into the Transpose 2-D Array function and into the array indicator. Since we are only reading the array values, not changing them, wire from the left to the right case structure tunnels, and then create a wire branch for the displays.

We don't want to end the program in the Graph state, so add a False Boolean constant and wire it to the output tunnel and to the While Loop conditional terminal. The easiest way is to right-click on the tunnel and choose Create >> Constant from the shortcut menu.

The program should execute the Analyze State next, so add an enumerated constant with value set to Analyze and wire it to the state-transition shift register. To complete the Analyze State as shown in Figure 6.30, copy the appropriate code from the Climate Simulation program.

FIGURE 6.29
Graph State

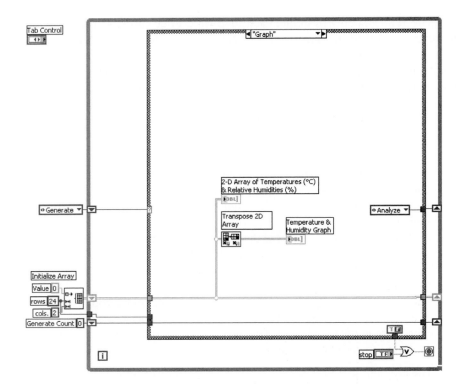

FIGURE 6.30
Analyze State

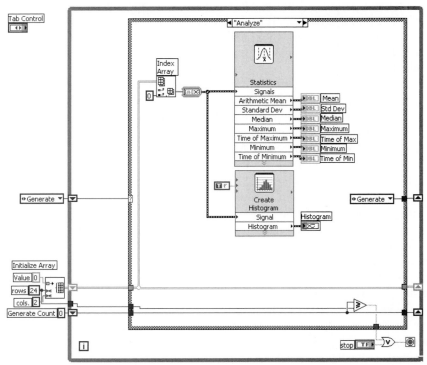

We will only analyze temperature data, as we did in the Climate Simulation program, so wire from the 2-D array Shift Register into the Index Array Function to separate column 0, the temperature values, from the array. Since we aren't changing the array values, just analyzing and displaying them, wire from the left to the right case structure tunnels, and then create a branch to the Index Array function.

We want to end the program in the Analyze State, so compare Generate Count to 24 and wire it to the output tunnel instead of the Boolean constant used in the Graph State.

The program should return to the Generate State, so add an enumerated constant with value set to Generate and wire it to the State-transition Shift Register as shown in Figure 6.30. Save, run, and test the program.

MATRICES

The preceding example program simulated climate measurement data, displayed it, and performed statistical analysis. There are other ways we can work with data structures in LabVIEW. For example, engineers and scientists frequently use linear equations to solve problems. Linear equations are those of the form

$$Y = mx + b$$

The equation can be graphed as a straight line with b as the Y intercept and m as the slope. A linear equation with three variables, x, y, and z represents a plane in 3-D space. More than three variables is a hyperplane.

Frequently we read to solve a series of linear equations, and determine the values of the variables. If we have m equations and n unknowns, and $m < n$, the system is underspecified and we can't find a unique solution. If $m = n$, a unique solution exists if the equations don't represent parallel hyperplanes. If $m > n$, the system is overspecified and doesn't have a unique solution.

We can solve a series of linear equations graphically or by Gaussian elimination. The graphical solution relies on intersecting lines or planes. Gaussian elimination reduces the number of variables to one, through adding scaled modifications of the first equation to the other equations in the system, solving for the value of the final variable, and back substituting the value into the equations to determine the values of the remaining variables.

We can implement Gaussian elimination on computers by representing the system of equations in matrix form:

$$Ax = b$$

where A is an $n \cdot n$ matrix, b is a given vector consisting of n elements, and x is the unknown solution vector to be determined.

Because matrices are useful for many applications, LabVIEW provides several functions and VIs for matrix analysis. The fundamentals of matrix analysis in LabVIEW are described in Help >> Index >> matrices.

Use the matrix data type instead of a 2-D array to represent matrix data because the matrix data type stores rows or columns of real or complex scalar data for matrix operations, particularly some linear algebra operations. The Mathematics VIs and Functions that perform matrix operations accept the matrix data type and return matrix results, which enables subsequent polymorphic VIs and functions in the data flow to perform matrix-specific operations. If a Mathematics VI or function does not perform matrix operations but accepts a matrix data type, the VI or function automatically converts the matrix data type to a 2-D array. If you wire a 2-D array to a VI or function that performs matrix operations by default, the VI or function automatically converts the 2-D array to a real or complex matrix, depending on the data type of the 2-D array.

A matrix is represented in LabVIEW with m rows and n columns using the same indexing notation as 2-D arrays where i represents the row and j represents the column in the element $a_{i,j}$:

$$A = \begin{bmatrix} a_{0,0} & a_{0,1} & \cdots & a_{0,n-1} \\ a_{0,0} & a_{1,1} & \cdots & a_{1,n-1} \\ \cdots & \cdots & \cdots & \cdots \\ a_{m-1,0} & a_{m-1,1} & \cdots & a_{m-1,n-1} \end{bmatrix}$$

There are several types of matrices:

Column Vector: An $m \cdot 1$ matrix—m rows and one column

Row Vector: A $1 \cdot n$ matrix—one row and n columns

Diagonal Matrix = All the elements other than the diagonal elements are zero

I = the Identity Matrix = A diagonal matrix with all the diagonal elements equal to one

U = Upper Triangular Matrix = All the elements below the main diagonal are zero

L = Lower Triangular Matrix = All the elements above the main diagonal are zero

Real Matrix = All the elements are real numbers

Complex Matrix = One or more of the elements is a complex number

Implementing matrix solutions sometimes requires that we transpose a matrix—that is, we interchange its rows and columns. If the matrix B represents the transpose of A, denoted by A^T, then $b_{j,i} = a_{i,j}$. Develop the Matrix Transpose program shown in Figure 6.31 by opening a new VI, adding the Matrix Transpose VI (Functions >> Mathematics >> Linear Algebra) to the block diagram window, and right-clicking the

FIGURE 6.31
Matrix Transpose program

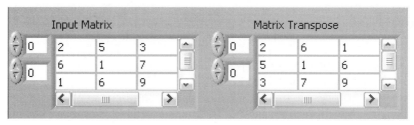

input and output terminal to add a control and an indicator on the front panel window. Then enter the values shown on the input. To avoid reentering the values the next time, we open and run the program, right-click the control border and change the default values (Data Operations >> Make Current Value Default). Save the program and run it.

We might also need to compute the determinant value from the values of the elements in the matrix. The determinant of a 2×2 matrix is

$$|A| = a_{1,1}a_{2,2} - a_{2,1}a_{1,2}$$

The determinant of larger square matrices is formed by summing the determinants of its elements. For example, the determinant of a 3×3 matrix is

$$|A| = a_{1,1}a_{2,2}a_{3,3} + a_{1,2}a_{2,3}a_{3,1} + a_{1,3}a_{2,1}a_{3,2} - a_{3,1}a_{2,2}a_{1,3} - a_{3,2}a_{2,3}a_{1,1} - a_{3,3}a_{2,1}a_{1,2}$$

Modify the Matrix Transpose program and save it as the Matrix Determinant program, as shown in Figure 6.32, using the Determinant VI (Functions >> Mathematics >> Linear Algebra) to compute the determinant of a matrix.

Other common operations are between matrices and scalars: for example, multiplying a matrix by a scalar, $C = \alpha A$. These use the arithmetic functions, and, for example, multiply all of the elements by the scalar, $c_{i,j} = \alpha * a_{i,j}$. These functions are polymorphic and will accept a matrix data type and a double-precision floating-point data type. Build the VI shown in Figure 6.33 to multiply a matrix and a scalar.

We can also perform arithmetical operations with matrices. For example, two (or more) matrices can be added or subtracted if they have the same number of rows and columns and $c_{i,j} = a_{i,j} \pm b_{i,j}$. Build the Matrix Addition program shown in Figure 6.34.

Two or more matrices can be multiplied if the number of columns of the first equals the number of rows of the second. If A has m rows and n columns and B has n rows and p columns, then $C = AB$ is an $m \times p$ matrix, where $c_{i,j} = \Sigma \, a_{i,k} b_{k,j}$. Build the Matrix Multiplication VI shown in Figure 6.35, using the A X B VI (Functions >> Mathematics >> Linear Algebra).

FIGURE 6.32
Matrix Determinant program

FIGURE 6.33
Matrix Scalar
Multiplication
program

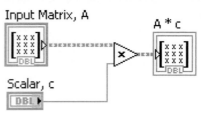

FIGURE 6.34
Matrix Addition
program

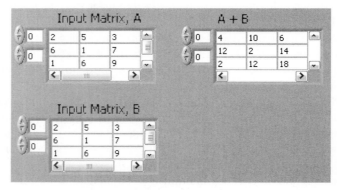

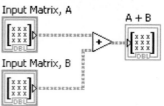

We can use the Solve Linear Equations VI to solve a set of equations such as the following:

$$3x + 2y - z = 10$$
$$-x + 3y + 2z = 5$$
$$x - y - z = 1$$

We express them in the form $Ax = b$, where A is the input matrix and b is the known vector.

$$A = \begin{bmatrix} 3 & 2 & -1 \\ -1 & 3 & 2 \\ 1 & -1 & -1 \end{bmatrix} \qquad b = \begin{bmatrix} 10 \\ 5 \\ -1 \end{bmatrix} \qquad x = \begin{bmatrix} x \\ y \\ z \end{bmatrix}$$

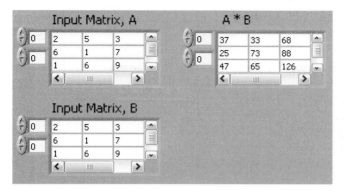

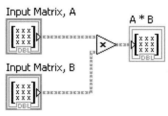

FIGURE 6.35
Matrix Multiplication VI

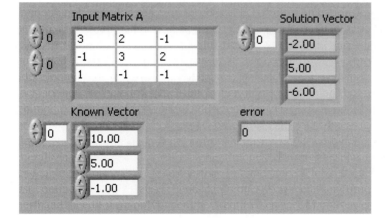

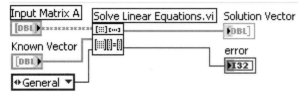

FIGURE 6.36
Linear Equation
Solver program

A Linear Equation Solver program as shown in Figure 6.36 will solve this and other systems of linear equations.

The Linear Algebra Calculator example (Help >> Find Examples) will solve systems of linear equations and perform many other matrix analyses. Locate and open it. Use it to solve the preceding system.

CLUSTERS

Clusters are very similar to arrays, but they can group elements of mixed data types, whereas arrays are restricted to a single type. For example, we can have an array of numeric values or an array of Boolean types, but we can't have an array with both numeric and Boolean types. However, we can bundle those two data types into a cluster.

A cluster is similar to a record or a struct in text-based programming languages. Bundling several data elements into clusters eliminates wire clutter on the block diagram and reduces the number of connector pane terminals needed by subprograms. Both arrays and clusters are either *all* controls or *all* indicators. Neither arrays nor clusters may have a mixture of controls and indicators.

Although cluster and array elements are both ordered, we must unbundle all cluster elements at once rather than index one element at a time, but we can use the Unbundle By Name function to access specific cluster elements. Clusters also differ from arrays in that they are a fixed size.

Most clusters on the Block Diagram have a pink wire color and data type terminal. Clusters of numeric values, sometimes referred to as points, have a brown wire color and data type terminal. We can wire brown numeric clusters to numeric functions, such as Add or Square Root, to perform the same operation simultaneously on all elements of the cluster.

Cluster elements have a logical order unrelated to their position in the shell. If you are programming in LabVIEW while reading this text, you will notice that the cluster is element 0, the second is element 1, and so on. If you delete an element, the order adjusts automatically. The cluster order determines the order in which the elements appear as terminals on the Bundle and Unbundle functions on the block diagram window. We can view and modify the cluster order by right-clicking the cluster border and selecting Reorder Controls In Cluster from the shortcut menu.

Use the Cluster functions to create and manipulate clusters. For example,

> ➤ extract individual data elements from a cluster.
> ➤ add individual data elements to a cluster.
> ➤ break a cluster into its individual data elements.

PICTURE INDICATOR

Arrays and clusters are useful for enhancing the graphical user interface with pictures. Picture functions and VIs use the cluster data type in drawing shapes, points, lines, and pixmaps into a picture indicator. Pixmaps are 2-D arrays of color, where each value corresponds to a color or to an index into an array of RGB (red, green, blue) color values, depending on the color depth.

Build the program shown in Figure 6.37 using the Draw Rectangle VI (Functions >> Graphics and Sound >> Picture) to draw two overlapping rectangles.

The rect input is a cluster that contains four numeric values that describe the left, top, right, and bottom coordinates of the outer edges of the rectangle. Horizontal coordinates increase to the right, and vertical coordinates increase to the

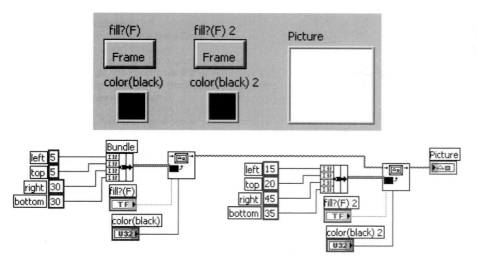

FIGURE 6.37
Draw Rectangles
program

bottom. Consequently, the left value is the horizontal coordinate of the left edge of the rectangle, the top value is the vertical coordinate of the top edge of the rectangle, the right value is the horizontal coordinate of the right edge of the rectangle, and the bottom value is the vertical coordinate of the bottom edge of the rectangle.

Right-click on the fill and color input terminals of the Draw Rectangle VI and create controls on the front panel window to give the user the option of filling and changing the color of the rectangles. Experiment with the location and dimensions of the rectangles. Try moving the rectangles by using a delayed loop.

Customize the rectangles with the Pen Input Terminal Cluster. It sets the width and style of the virtual pen the program uses to draw the picture. Width sets the width in pixels, and style sets the style:

0 Solid (default)

1 Dash

2 Dot

3 DashDot

4 DashDotDot

SUMMARY

> Array and cluster data structures facilitate access to individual elements of data as well as to the entire set.

> The array structure associates a name to a block of memory containing the data and associates all the individual values with the block.

> An array consists of an identifier, elements, dimensions, and indexes. The identifier is the label.

> LabVIEW arrays can be multidimensional.

> It takes two steps to define the array from the front panel window: add an Array shell and drag a data object or element into the shell.

➤ Array elements can be numeric, Boolean, string, path, refnum, or cluster data types, but we can't mix the types in the array.

➤ Zero-based indexes denote the position of an element in an array. Normally, the indexes are listed in (column, row) order.

➤ The For Loop repetition structure repeats N times and exits where N is the loop-count terminal.

➤ We can program equations into VIs with a Formula Node that contains one or more C-like statements delimited by semicolons.

➤ We can create arrays by wiring values through indexed-enabled loop tunnels.

➤ To graph data in LabVIEW, we add a graph indicator on the front panel window and wire the data to its block diagram icon.

➤ There are several types of graphs (e.g., waveform, XY, 3-D) and charts available in LabVIEW that can be customized for many applications.

➤ LabVIEW includes many functions for working with arrays that return values for maximum, minimum, size. They also divide arrays into subsets, concatenate, initialize, insert elements, replace subsets, reshape, transpose, rotate, sort, split, and thus perform other functions.

➤ We can use LabVIEW VIs to generate many signals, including sinusoids, sawtooth, square, and other waveform shapes. VIs can also generate random, uniform, and Gaussian noise to make simulations more realistic.

➤ LabVIEW contains a large number of statistical analysis functions and VIs, including maximum, minimum, mean, mode, median, standard deviation, variance, RMS, and others.

➤ We can represent matrices in LabVIEW with the matrix data type.

➤ LabVIEW contains a large number of VIs and functions for matrix analysis, including transpose, determinant, scalar multiplication, matrix addition, matrix multiplication, linear equation solver, and others.

➤ Clusters are very similar to arrays, but they can group data elements of mixed types, whereas arrays are restricted to a single type. You can't mix controls and indicators in an array or in a cluster.

➤ Picture Functions VIs use the cluster data type in drawing shapes, points, lines, and pixmaps into a picture indicator.

➤ LabVIEW allows mixing data types in many functions coercing one of the data types. This should be avoided, especially when working with large arrays as LabVIEW copies the memory in this process.

EXERCISES

1. Read the chapter and build and test the Temperature Array PtByPt program.

2. Build and test the Temperature and Humidity Array PtByPt program.

3. Build and test the Temperature and Humidity Array PtByPt with Noise and Stats program.

4. Update your notebook with new information from the chapter, including the shortcut keys and data types.

5. Change the Temperature and Humidity Array PtByPt with Noise and Stats to a state machine. Save it as the Climate Simulation State Machine. Use a Type Defined Control to create the state-transition enumerated constants. Save the control.

6. Modify the Climate Simulation State Machine by placing the simulation equations and formula node with noise in a sub VI.

7. Modify the Climate Simulation State Machine by replacing the formula node with a LabVIEW point-by-point sine function.

8. Add code to analyze the statistics for the humidity data in the Climate Simulation State Machine.

Display the results on a third page of the tab control.

9. Compare the statistics in the Climate Simulation State Machine with and without noise.

10. Use different means of introducing noise in the Climate Simulation State Machine. Experiment with Gaussian noise, a Random Number function, and Periodic Random noise.

11. The Climate Simulation State Machine generates data and noise point-by-point. Replace the point-by-point generation with waveform generation sub VIs from the Functions Pallette. Is a For Loop necessary with the waveform programs? _____

12. Modify the Climate Simulation State Machine to generate temperature and humidity data for a week, using nested For Loops. Save the program as Climate Simulation One Week State Machine.

13. Replace the simulation in the Climate Simulation State Machine with actual measurement. Use the temperature transducer for one channel and a sine wave from the function generator for the other. Save it as the Climate Measurement State Machine.

14. Open the Getting Started with LabVIEW pdf file. Work through Chapter 3, "Analyzing and Saving a Signal." Build and save the VIs.

15. Build a 1-D Array of 10 temperature measurements with a While Loop. After the array is built, reverse it and display before and after the reversal.

16. Build a 1-D Array of 10 temperature measurements with a While Loop. Update displays of the current value, the average value, the maximum value, and the minimum value as the loop is running.

17. Build a 1-D Array of 10 temperature measurements with a For Loop. After it is built, sort the array from highest to lowest. Display before and after sorting.

18. What is the difference between using a waveform chart and a waveform graph in the Temperature and Humidity Array PtByPt with Noise and Stats program? When should you use a graph instead of a chart?

_____

_____

19. Develop your own program that creates and manipulates array data.

20. Build a new Matrix Transpose program using the Transpose Array Function. Is this program different from the Matrix Transpose VI example developed in the text? Explain.

_____

_____

21. Experiment with the Linear Algebra Calculator VI example to complete some matrix analysis of interest.

22. Download the Tetris game from ni.com Home > NI Developer Zone > LabVIEW Zone. Extract the zip files to c:\temp\tetris. Play the game for a few minutes. Study the block diagram window, and answer these questions:

   a. What happens each time the While Loop iterates?

   _____

   b. Write the names of 5 states in the lower Case Structure below.

   _____

   _____

   c. Which shift register (counting down from the top on the right side of the While Loop) contains the score?

   d. What functions from the Functions >> Programming >> Numeric subpalette are used? Write their names and symbols below.

   _____

   _____

   e. What functions from the Functions >> Programming >> Comparison subpalette are used? Write their names and symbols below.

   _____

   _____

f. Explain the process of using the Array Size function and the Array Index function on the 1-D array in the lower left of the Tetris block diagram window that selects the state for moving the block.

———————————————————————

———————————————————————

23. Build a program that displays the Tetris Shapes using the same process as the Tetris game.

24. Modify the Draw Rectangles program to allow the user to select the size and location of the rectangles. Increase the size of the picture indicator on the front panel window.

25. Modify the Draw Rectangles program to use a loop to move the rectangles around on the screen and change their size and color automatically while the program is running.

26. Create a program that uses the picture indicator to do something that is interesting to you.

Input and Output

INTRODUCTION

This entire text is focused on input and output, but up to this point, our DAQ applications have used Express VIs that don't require knowledge of the fundamental process. While they are easy to program, Express VIs aren't as versatile as lower-level functions. This chapter explains the fundamental paradigm of opening a resource, reading or writing to it, closing it, and handling errors. It can be applied to a variety of input/output data exchanges such as reading and writing files, Internet communications, data acquisition, and communications with external instruments.

Computers are useful when they interact with the world through input and output (I/O) devices. So far we have provided input from the mouse and keyboard and received output on the monitor. There are many other ways to receive input and to send output. For example, the computer can receive input from a microphone, a joystick, the Internet, sensors, local networks, scanners, and so on. Similarly, computers can send output to the Internet, local networks, instruments, sensors, motors, printers, and so on.

The computer operating system controls the flow of information between devices. Our programs work with the operating system to request communications activities between devices. Our program might cause a stream of bytes to be sent from an input device such as a joystick to an output device such as a monitor. Or our program might generate the stream of bytes, store it in memory, and send it to an output device such as a hard drive. So our program controls the content and the flow of the information. We will apply the concepts presented in this chapter to developing more efficient and versatile DAQ applications in Chapter 8.

OUTLINE

FILE I/O

One of the most common I/O tasks is reading and writing information from and to files. We have already used the Write to Measurement File and the Export Waveforms to Spreadsheet File VIs. In this chapter, we will learn more efficient and versatile file I/O.

Computers store data in binary format, 0s and 1s, or bits. LabVIEW functions and VIs convert binary data to the characters we prefer. A common format used by engineers and scientists is a spreadsheet format, which displays a 2-D array in columns and rows using the tab character to separate columns and an end-of-line character to separate rows. Again, LabVIEW will do this for us in high-level VIs, like the ones we have used so far, or allow us to control the formatting in lower-level VIs and functions.

For an example, we will quickly modify the Temperature & Humidity program developed previously to save the 2-D array of simulated temperature and humidity values to a spreadsheet file. All we have to do is add the Write To Measurement File VI (Functions >> Programming >> File I/O), as shown in Figure 7.1 and configure it as shown in Figure 7.2.

Using the configuration in Figure 7.2, the file will contain the information shown in Figure 7.3. The file was opened in a spreadsheet program from the path shown in the File Name box in Figure 7.2. We do not have to enter a path in this box for the Write To Measurement File VI to execute. We could also wire a control or constant to the File Name input to the Express VI on the block diagram window.

We chose Save to one file in the configuration dialog, so LabVIEW saves all the data to one file. We could also check the box asks users to select a file. It will pop up a dialog when the Express VI executes which is an alternative to using the path typed

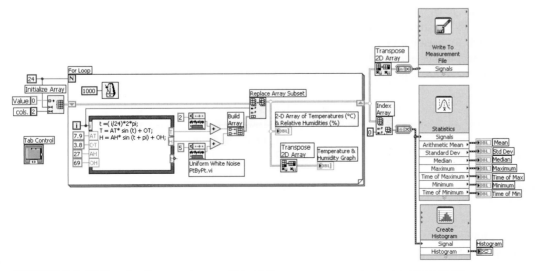

FIGURE 7.1 Writing the Temperature and Humidity program array to a spreadsheet file

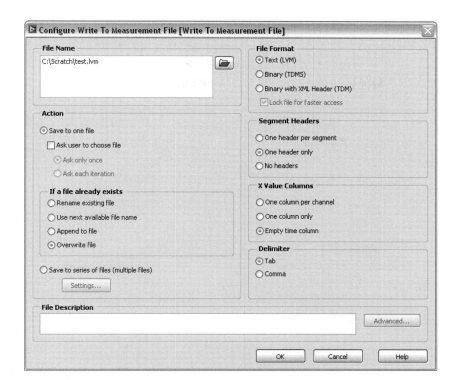

FIGURE 7.2 Write to Measurement File configuration

LabVIEW Measurement			
Writer_Version	0.92		
Reader_Version	1		
Separator	Tab		
Multi_Headings	No		
X_Columns	Multi		
Time_Pref	Relative		
Operator	rking		
Date	1/5/2008		
Time	35:02.4		
End_of_Header			
Channels	2		
Samples	24		24
Date	1/5/2008		1/5/2008
Time	35:02.4		35:02.4
X_Dimension	Time		Time
X0	0.00E+00		0.00E+00
Delta_X	1		1
End_of_Header			

FIGURE 7.3 Two-dimensional Temperature & Humidity array displayed in Microsoft's Excel® spreadsheet software

(Continued)

FIGURE 7.3
(Continued)

X_Value	Untitled	X_Value	Untitled 1
0	4.422626	0	72.87974
1	5.789893	1	60.18888
2	7.854254	2	58.197714
3	9.038018	3	44.930404
4	10.499385	4	41.287626
5	11.522397	5	44.809484
6	11.600237	6	38.894296
7	13.250408	7	45.082713
8	8.749913	8	43.45629
9	7.508309	9	48.320387
10	6.927751	10	60.202933
11	7.711296	11	64.257169
12	2.6779	12	67.679098
13	1.963446	13	72.525866
14	−1.226467	14	85.219632
15	−2.758034	15	86.752739
16	−3.701376	16	95.570317
17	−5.369056	17	92.666683
18	−4.156853	18	97.849376
19	−4.333894	19	95.062218
20	−3.525275	20	95.340923
21	0.201054	21	91.45569
22	1.72508	22	78.762098
23	3.030452	23	73.154134

into the File Name box. The Express VI executes only once in our application, but it could execute multiple times in other applications, for example if it is in a loop. The dialog box gives the user the option of selecting a file each time the Express VI runs if you place a checkmark in the Ask user to choose file checkbox.

There may be a file existing at the same name and in the same path as the one specified by the user. If so, we have several options:

➢ Rename existing file—Renames the existing file if Reset is TRUE. Reset is an input terminal to the Express VI that will accept a value from a Boolean constant or control.
➢ Use next available name—Appends the next sequential number to the filename if Reset is TRUE. For example, if test.lvm exists, LabVIEW saves the file as testl.lvm.
➢ Append to file—Appends the data to the existing file.
➢ Overwrite file—Replaces data in an existing file if Reset is TRUE.

We can also choose to save to series of files. If we choose to save to a series of files, the Settings button will become active allowing us to open the Configure Multi-file Settings dialog box to configure the files.

We can choose one of three file formats. The Text (LVM) format is a text-based file that can be read by many other programs like spreadsheet and word processing software (LVM-LabVIEW Measurement). The file extension is .lvm. The binary (TDMS) file is more efficient in speed and size, but binary formatted files can't be read by other programs without writing a separate translator program in LabVIEW. The binary with XML Header (TDM) has the same advantages and disadvantages as the previous binary format; however the XML header information facilitates searching in a data-base manner.

Headers contain information about the data, or meta data. The header information is contained in the dynamic data type used by Express VIs. The header is divided into two sections in Figure 7.3. The first section contains information about the Write to Measurement File configuration and the second section contains information about the data. We can configure the Express VI to write one header per segment, one header only, or no headers.

The X value column of our data is the row index (0-24) of the two-dimensional array. We have the options of creating a separate column for the time data that each channel generates. This option includes a column of values from the x-axis for every column of values from the y-axis. We would select this option if we had acquired signals using the DAQ Assistant of different types or at different acquisition rates rather than generating simulated data. We could also choose "One column only." This option creates only one column for the time data for signals acquired in the DAQ Assistant at the same acquisition rate. We configured for an empty time column in Figure 7.2 since we are simulating data and not measuring time and signals with the DAQ Assistant.

A delimiter can be placed between data items in the file to facilitate reading with other software. We configured for the Tab delimiter since it is commonly used by spreadsheet software. We could also choose comma delimiters.

We can write a description of the data in File Description box. LabVIEW appends the text you enter in this text box to the header of the file. This is a very useful facility if we are conducting a series of experiments and want to document the differences between the experiments.

LabVIEW provided us a very quick way to perform a complex task with the Write to Measurement File VI. However, there is considerable overhead associated with the Express VIs as you can see by opening the Express VI front panel window and using the VI Heirarchy tool.

We can build a more versatile and efficient VI with lower-level file VIs, and they are easy to use. We replace the Express VI with three steps:—open a file, write or read to it, and close the file, as shown in Figure 7.4.

We open a file with the Open/Create/Replace File VI (Functions >> Programming >> File I/O), which has several input terminals to configure the open operation. Figure 7.4 shows a control of file-path data type control connected to the file-path input terminal, allowing the user to specify the path and name of the file from the GUI. If the user leaves the file path control empty (default) or inputs <Not A Path>,

FIGURE 7.4 File I/O sequence

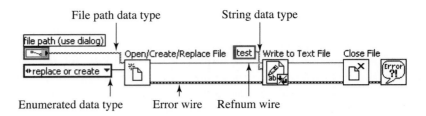

File path data type · String data type · File path (use dialog) · Open/Create/Replace File · test · Write to Text File · Close File · replace or create ▼ · Enumerated data type · Error wire · Refnum wire

the VI will display a dialog box when the user runs the VI that halts the program while the user selects a file name and location.

The replace or create input specifies the way files are opened with an enumerated constant or control:

0 Opens an existing file (default)

1 Opens an existing file or creates a new file if one does not exist

2 Creates a new file or replaces a file if it exists and the user gives permission

3 Creates a new file

4 Opens an existing file for read only

The Open/Create/Replace file VI creates a reference number, or refnum, which is a unique identifier for an object. Because a refnum is a temporary pointer to an open object, it is valid only for the period during which the object is open. LabVIEW remembers information associated with each refnum, such as the current location for reading from or writing to the object and the degree of user access, so you can perform concurrent but independent operations on a single object. The refnum output wire in Figure 7.4 connects the open, write, and close file functions.

Spreadsheet programs, and many other programs, require the data in the file to be in American Standard Code for Information Interchange (ASCII) format. These ASCII files are also known as text files, character files, or string files. The ASCII character set represents the letters and most punctuation marks used in the English language, for a total of 128 characters. Using the ASCII character set might not be sufficient in all applications because many languages require characters not found in ASCII, so alternatives such as ISO Latin-1, also known as ISO-8859-1, which describes characters used in most western European languages, and the Unicode Standard that encodes most of the world's characters and symbols (www.unicode.org) should be considered.

This text uses ASCII files, but LabVIEW can also read and write binary and datalog files. Use binary files for fast I/O, higher precision, and smaller files. Use datalog files when saving several data types in a record. Use text files to access the data easily with other applications. Refer to Help >> Index >> File I/O for more information.

The Write to Text File VI requires the input information to be in string data type. The Format into String and the Array to Spreadsheet String functions can convert numbers to a string, as shown in Figure 7.5 example that writes a random number to a text file.

The number-to-string conversion functions use a format-string input, a code specifying how to convert numeric data to or from a string. LabVIEW allows the user

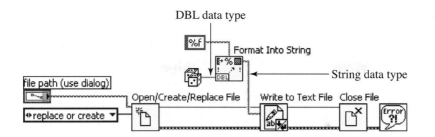

FIGURE 7.5
Converting numeric
data to ASCII data
before Writing to File

to specify a wide variety of formats; consequently, formatting can be very tedious. Search LabVIEW Help for Format Specifiers >> Syntax Elements for more details on formatting. Table 7.1 shows a few examples including the %f, floating point, format used in the Figure 7.5 example VI.

TABLE 7.1   Format Examples (LabVIEW Help>Format Specifiers Syntax Elements)

Type	Argument(s)	Format String	Resulting String
Automatic formatting	12.00	%#g	12
(%g)	12000000	%#g	1.2E+ 6
Decimal (%f)	12.67	score= %.0f%%	score= 13%
Floating point (%f)	12.67	Temp: %5.1f	Temp: 12.7
	12.67 N	%5.3f	12.670 N
	12.67 N	%5.3{mN}f	1267.000 mN
	12.67 N	%5.3{kg}f	12.670 ?kg
Scientific/engineering	12.67	%.3e	1.267E+ 1
notation (%e)	12.67	%^.3e	12.670E+ 0
SI notation (%p)	12000000	%.2p	12.00M
	12000000	%_2p	12M
Hexadecimal (%x)	12	%-02x	0A
Octal (%o)	12	%-06o	000012
Binary (%b)	12	%b	1100
Relative time (%t)	91.80	%.2t	01:31.80
		% < Hours:%H	
	91.8	Minutes:%M	Hours:00 Minutes:01
		Seconds:%S > t	Seconds:31
Absolute time (%T)	2001	% < %.3X%x > T	12:00:00.000 AM
			01/01/2001
	2001	% < %Y.%m.%d > T	2001.01.01
String (%s)	Smith John	Name: %s, %s.	Name: Smith, John.
	Hello, World	String: %10.6s	String:___Hello,

The underline character (_) represents spaces.

FIGURE 7.6 Edit
Format String dialog
box

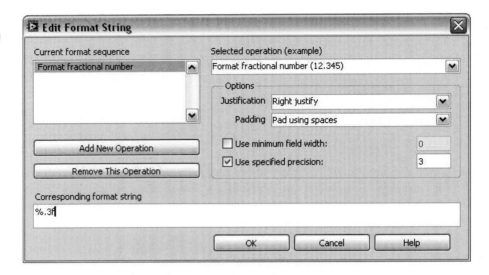

The Edit Format String Dialog Box can help set up formats. However, it is available only from the Format Into File or the Format Into String functions. If your program doesn't require one of those functions, place one on the block diagram window anyway, figure out the format, and delete the function but keep the format string constant. Right-click the format-string terminal and select Edit Format String from the shortcut menu to display the Edit Format String dialog box (Figure 7.6). Click the Help button on the bottom to obtain more information.

ERROR HANDLING

Two wires continue from the Open File Function through the program to the Close File Function, as shown in Figure 7.4 and Figure 7.5.

The wire connecting the upper terminals is the file reference number (refnum). The wire connecting the lower terminals is the error cluster wire. If an error occurred, this VI returns a description of the error and optionally displays a dialog box, shown in Figure 7.7.

FIGURE 7.7 Example error dialog

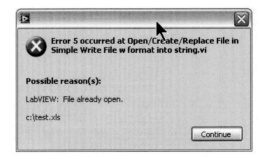

Error checking in a program provides information on why and where errors occur. We can use this information to handle errors in appropriate ways that make our programs more robust. I/O (file, serial, instrumentation, data acquisition, and communication) has the possibility to cause several errors; for example, mistyping a file path. Additional errors can be caused by the following:

> Initializing communications incorrectly
> Writing improper data to an external device
> An external device lost power, is broken, or is working improperly
> Operating system software upgrade changed the path to a file or the functionality of a VI or library

By default, LabVIEW automatically handles errors by suspending execution. However, we can choose other error handling methods. For example, if an I/O VI on the block diagram window programatically times out, we might not want the entire application to stop. We might want the VI to retry for a certain period of time. In LabVIEW, we can make these error handling decisions by connecting error wires between nodes. Whenever we connect error wires, we suspend automatic, error handling.

VIs and functions return errors in one of two ways—with numeric error codes or with an error cluster. Typically, functions use numeric error codes, and VIs use an error cluster, usually with error input and output terminals. The error-in and error-out cluster wires contain three parts:

1. Status is a Boolean value that reports TRUE if an error occurred.
2. Code is a 32-bit signed integer that identifies the error numerically. A nonzero error code coupled with a status of FALSE signals a warning rather than a fatal error. Select Help>>Explain Error to display an error code description in the Explain Error dialog box. Enter the error code number in one of the Code fields and click the Status button to change the button to a red X. A description of the error code appears in the Explanation field. Help>> Search LabVIEW Help contains a list, General LabVIEW Error Codes. For example, if we enter "error 5" in the search box and select General LabVIEW Error Codes in the search results, we will see that error 5 is a file already open, the error shown in Figure 7.7.
3. Source is a string that identifies where the error occurred. The error occurred when the Open/Create/Replace File VI executed in the Figure 7.7 example.

Error handling in LabVIEW follows the data-flow model. We wire the error information from the beginning of the program to the end and include an error handler VI at the end of the program to determine if the program ran without errors, as shown in Figure 7.8. The Simple Error Handler Function (Functions >> Dialog & User Interface palette) pops up a dialog box if an error occurs.

As the error information flows through the VI, LabVIEW tests for errors at each execution node. If LabVIEW does not find any errors, the node executes normally. If LabVIEW detects an error, the node passes the error to the next node without executing that part of the code. The next node does the same thing, and so on. At the end of the execution flow, LabVIEW reports the error.

Error wires also create an execution sequence. We don't want the Write to Text File function, in the Figure 7.8 example VI, to execute before the file has been

FIGURE 7.8 Error handling

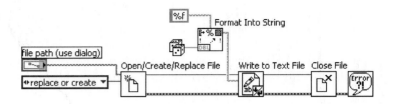

opened. Wiring the error data cluster from the Open/Create/Replace function to the Write to Text File function uses the LabVIEW data flow paradigm to make the Write to Text File function wait until the Open/Create/Replace File function completes execution and provides the error data.

When you wire an error cluster to the selector terminal of a Case Structure, as shown in Figure 7.9, the case selector label displays two cases, Error and No Error. If you are coding in LabVIEW while studying this text, you will notice that the border of the Case structure changes color—red for Error and green for No Error. If an error occurs, the Case Structure executes the Error Case. This gives the programmer considerable flexibility in handling errors. We can write programs where

1. Any error aborts the program.
2. The program ignores all errors.
3. The program evaluates the value of the error code and/or the source of the error.
   a. Open a dialog box announcing an error and ask the user for input.
   b. Abort execution of certain parts of the program and allow others to continue.
   c. Delete the error.
   d. Change control flow to handle the error in the program and continue execution.

FIGURE 7.9 Error and No Error Case Structure

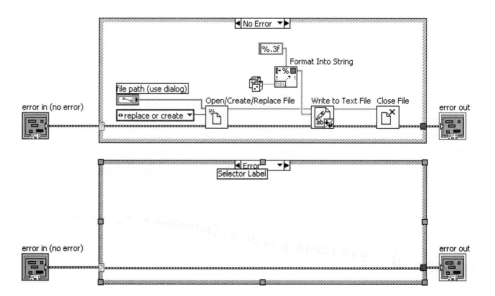

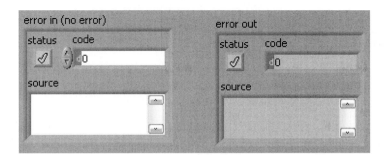

FIGURE 7.10 Error control and indicator for wiring error cluster through sequential programs

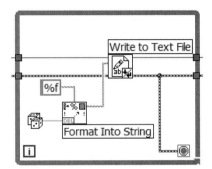

FIGURE 7.11 Stopping a while loop if an error occurs

Error Control and Indicators are available from the Controls >> Array, Matrix, & Cluster Subpalette, as shown in Figure 7.10. They can be used to add error-in and error-out terminals to subprogram connector panes. It is conventional to add these terminals on the lower left and lower right of all sub VI connector panes.

You can wire an error cluster to the conditional terminal of a While Loop as shown in Figure 7.11 to stop the iteration of the While Loop if there is an error. You can use the unbundle function to extract the value of Status in the error cluster and wire it to an Or function along with the stop control. Wire the output of the Or function to the conditional terminal configured to Stop if True. Consequently, the While Loop will stop if the Stop button is pressed or if an error occurs.

DISK STREAMING

We might wish to save each data point after it is generated so we don't lose data if the system crashes before a loop exits. We might also wish to be more efficient and only open the file once, write to it many times, and close it once. This will reduce delays in a looping action where time is critical, for example, in high-speed sampling.

Figure 7.12 shows an example of disk streaming where two random numbers are generated and written to two columns in a spreadsheet file. Column headers are written before the loop executes. A tab constant is used to separate the numbers into two columns and an end-of-line (EOL) constant designates the end of each row.

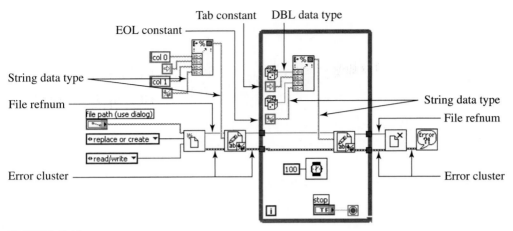

FIGURE 7.12 Disk-streaming example

We can use disk streaming in the state machine architecture. A state machine isn't required for disk streaming, but saving data is usually one of several states in larger programs, so it is a common practice. For example, we could replace the Write to Measurement File Express VI in the Temperature Measurement State Machine, as shown in Figure 7.13. The Convert from Dynamic Data Express VI (Functions >> Programming >> Express >> Signal Manipulation) converts the Dynamic Data type to a one-dimensional array of DBL type. The Array to Spreadsheet String Function (Functions >> Programming >> String) converts the data to ASCII format, inserts tabs between columns, and inserts EOL characters at the end of each row of data. The %.3f format string specifier will cause the data to be saved as floating point with three decimal places.

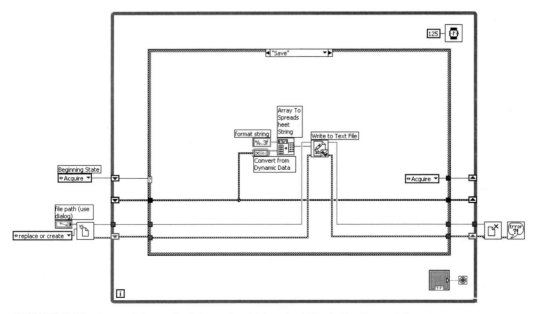

FIGURE 7.13 Convert dynamic data and write to a text File in the Save state

This example saves only the amplitude values from the dynamic data type. Because the program doesn't change the data values, just saves them, we wire from the left to the right case structure tunnels, and then create a branch to the convert function.

If we place the Open/Create/Replace File function (Functions >> File I/O) outside of the While Loop as shown in Figure 7.13, it will execute only once before the while loop. It could also be placed in an initialize state.

After the While Loop has saved 24 lines and exits, the program should close the file, so add a Close File function (Functions >> File I/O) as shown. Wire the refnum and error wires from the Write to Text File function, through the tunnels, to the Close File function.

Add the Simple Error Handler VI (Functions >> Timing) as shown. Wire the error terminal from the Close File function to the Simple Error Handler VI.

We can open the file with a spreadsheet or text program and compare the data with the array displayed on the front panel window. Close the file displayed in the spreadsheet or text program before executing the VI again.

DAQ I/O

We can apply the same paradigm to other I/O operations, such as data acquisition using the DAQmx VIs, as shown in Figure 7.14. The DAQmx VIs can replace the DAQ Assistant in the previously developed Climate Measurement State Machine or in the Temperature Measurement State Machine. The DAQmx VIs allow the user to control the same configuration information as in the DAQ Assistant. As we will explain in Chapter 8, we can improve efficiency with the DAQmx VIs to open a DAQ device and start acquisition before a loop, read the data repetitively in the loop, and clear the process and handle errors after the loop.

The steps to create an application with the DAQmx VIs are the same as those with the DAQ Assistant:

1. Configure the measurement hardware in MAX.
2. Create tasks and channels.
3. (Optional) Set timing.
4. (Optional) Set triggering.
5. Read or write data.
6. Clear.

The DAQmx VIs are polymorphic and can accept or return data of various types, such as scalar values, arrays, or waveforms.

The DAQmx Create Virtual Channel VI creates a virtual channel or set of virtual channels and adds them to a task. You can open or create tasks by using a DAQmx task name constant or control to select a task that you created and saved in the DAQ Assistant, or by using the DAQmx Create Virtual Channel VI or DAQmx Create Task VI to create a task programmatically. When you use a DAQmx task name constant or control to select a task, LabVIEW loads that task into memory one time only, even if the control or constant is inside a loop.

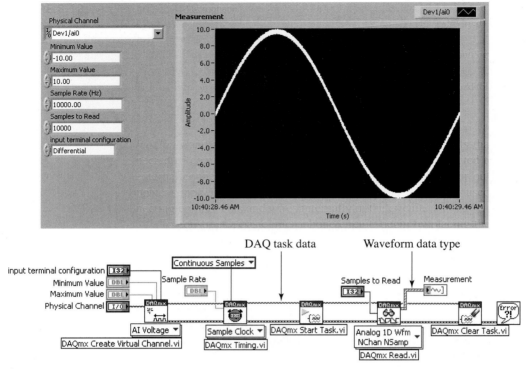

FIGURE 7.14 I/O with DAQmx VIs

Configure this polymorphic VI by selecting the I/O type of the channel with the polymorphic selector, as shown in Figure 7.15. The VI can be configured to instances similar to those in the DAQ Assistant, such as analog input, digital output, or counter output; the measurement or generation to perform, such as temperature measurement, voltage generation, or event counting; and in some cases, the sensor to use, such as a thermocouple or RTD for temperature measurements. It is best to place this VI before a loop, so it doesn't create a new task in each iteration of the loop.

The polymorphic DAQmx Timing VI sets the source of the Sample Clock, the rate of the Sample Clock, and the number of samples to acquire or generate. Rate is the sampling rate in samples per channel per second. Source specifies the source terminal of an external clock. Leave this input unwired to use the default onboard clock of the device (the PCI-6024E, for example). Sample mode specifies if the task acquires or generates samples continuously or if it acquires or generates a finite number of samples. The sample mode options include the following:

> Continuous Samples—Acquire or generate samples until the DAQmx Stop Task VI runs

> Finite Samples—Acquire or generate a finite number of samples

> Hardware Timed Single Point—Acquire or generate samples continuously using hardware timing without a buffer

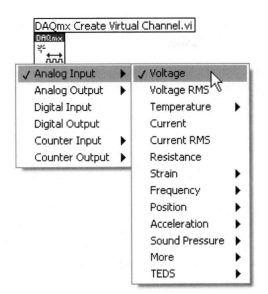

FIGURE 7.15 DAQmx Create Virtual Channel polymorphic selector

Samples per channel specifies the number of samples to acquire or generate for each channel in the task if sample mode is Finite Samples. If sample mode is Continuous Samples, NI-DAQmx uses this value to determine the buffer size. A buffer is memory allocated to the task. If your acquisition is finite, NI-DAQmx allocates a buffer equal in size to the value of the samples per channel. For example, if you specify samples per channel of 1000 samples and your application uses two channels, the buffer size would be 2000 samples. Thus, the buffer is exactly big enough to hold all the samples you want to acquire.

If the acquisition is continuous, NI-DAQmx allocates a buffer equal in size to the value of the samples per channel, unless that value is less than the value listed in Table 7.2.

Using very large buffers may result in diminished system performance due to excessive reading and writing between memory and the hard disk. Reducing the size of the buffer or adding more memory to the system can reduce the severity of these problems.

TABLE 7.2 DAQmx Buffer Sizes (LabVIEW Help>How is Buffer Size Determined?)

Sample Rate	Buffer Size
No rate specified	10 kS
0–100 S/s	1 kS
100–10,000 S/s	10 kS
10,000–1,000,000 S/s	100 kS
> 1,000,000 S/s	1 MS

The DAQmx Start Task VI transitions the task to the running state to begin the measurement or generation. If you do not use this VI, a measurement task starts automatically when the DAQmx Read VI runs. If you do not use the DAQmx Start Task VI and the DAQmx Stop Task VI when you use the DAQmx Read VI or the DAQmx Write VI multiple times, such as in a loop, the task starts and stops repeatedly. Starting and stopping a task repeatedly reduces the performance of the application.

The DAQmx Read VI reads samples from the task or virtual channels. The instances of this polymorphic VI specify what format of samples to return, whether to read a single sample or multiple samples at once, and whether to read from one or multiple channels. The DAQmx Read VI outputs data in waveform data type which includes an array of signal amplitude values, start time, and the time interval between samples.

The DAQmx VI clears the task from memory, releasing the memory for other uses. Before clearing, this VI stops the task, if necessary, and releases any resources the task reserved. You cannot use a task after you clear it unless you recreate the task.

When you use the DAQmx Create Virtual Channel VI and you do not specify a task to which to add the created channels, NI-DAQmx creates a new task and allocates resources for it. LabVIEW does not automatically free up these resources until the application completes.

INSTRUMENT CONTROL

Instrument control uses a computer and software to automate one or multiple instruments by sending commands and data between the instrument and the computer. Examples of instruments are digital multimeter (DMM), oscilloscope, arbitrary waveform/function generator, DC power supply, switch, power meter, spectrum analyzer, and RF signal generator.

The reasons we want to automate instrument control are similar to those for using data acquisition. For example, we are measuring impedance with an LCR Meter for one of our research projects. We need to measure five values at three frequencies, or 15 data points every 5 s for several hours. We automated the process with LabVIEW to avoid manually taking data that would entail watching the clock, recording the values, and typing them into a computer program. Consequently, there would be a delay as each data point was recorded, so all 15 wouldn't be read simultaneously. In addition, there might be typing and reading errors. Finally, recording 15 values every 5 s for several hours is not an enjoyable occupation. LabVIEW can automate the process by reading all of the values with small delay between each, providing nearly simultaneous measurements. Typing and reading errors will be eliminated. The program will automatically analyze the data, write it to files, and allow the researchers to do more enjoyable tasks.

To control an instrument, a LabVIEW program must send commands to the instrument in a language the instrument can understand, and LabVIEW must be able to understand information it gets back from the instrument. One way to accomplish this is to use special software called instrument drivers. Instrument drivers eliminate the need to learn complex, low-level programming commands. Not all instruments have instrument drivers for LabVIEW, but many are available on the ni.com instrument driver network.

Communication Busses

There are several ways to connect instruments to a computer. A very popular one, the General Purpose Interface Bus (GPIB), follows the IEEE 488 standard and provides digital, 8-bit parallel, byte-serial communications with data transfer rates of at about 1–2 Mbytes/s. The HS488 standard extension to GPIB has higher data rates (8 Mbytes/s).

GPIB controllers can be plugged in to a slot in the PC's PCI bus, they can also communicate with Ethernet, parallel, serial, USB, and FireWire interfaces. In these cases, the GPIB device both controls the communications and converts the data from one bus to another. A GPIB controller can communicate with up to 14 instruments, and the bus can be extended for more connections. Furthermore, you can plug more than one controller board into a PC. Whenever instruments communicate over a bus, there is the possibility of signal collisions, so the GPIB protocol gives control of the bus to one device at a time. Devices are categorized as controller, talker, or listener. The device requests service from the controller, receives control of the bus, transmits, and

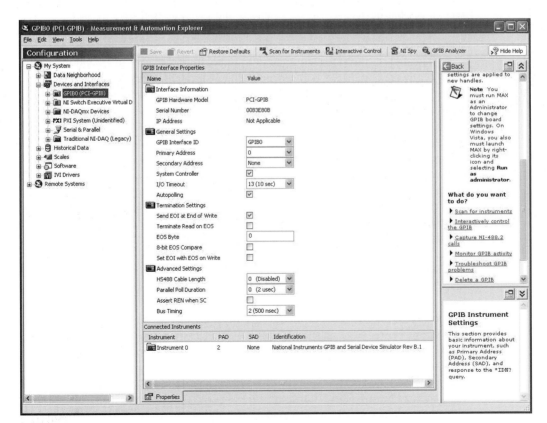

FIGURE 7.16 Recognizing a GPIB plug-in board and instrument in MAX

TABLE 7.3   IEEE 488.2 Control Sequences (from ni.com)

Keyword	Name	Compliance
RESET	Reset System	Mandatory
FINDRQS	Find Device Requesting Service	Optional
ALLSPOLL	Serial Poll All Devices	Mandatory
PASSCTL	Pass Control	Optional
REQUESTCTL	Request Control	Optional
FINDLSTN	Find Listeners	Optional
SETADD	Set Address	Optional, but requires FINDLSTN
TESTSYS	Self-Test System	Optional

informs the controller it has terminated transmission. Only one device can talk, but several can listen.

MAX contains NI-488 drivers, and it will recognize NI GPIB devices. Open MAX and expand the "Devices and Interfaces" subdirectory below "My System." If MAX recognizes the device, it will be visible in the configuration tree, as shown in Figure 7.16. If your device did not appear, use the GPIB troubleshooter in the MAX Help window.

If a cable is connected from the plug-in board to the instrument and the instrument is powered on, MAX will recognize the instrument. Click on "Scan For Instruments," and expand the tree below the GPIB board. If your GPIB device is compliant with the Standard Commands for Programmable Instruments (SCPI) standard, the name and address will appear in the "Connected Instruments" window.

After the instrument is recognized, LabVIEW can communicate with it. Table 7.3 presents controller messages in the IEEE 488.2 standard. Table 7.4 lists standard protocols that combine several commands to execute common operations. Table 7.5 lists some common standard commands for instrument communication.

You can also connect instruments with RS-232, RS-422, and RS-485 serial communication busses. Serial communication is slower than GPIB, as it transmits one bit at a time over a single line. Serial communication groups a series of bits that include a start bit, the data bits, a parity bit, and stop bits.

When using serial communication, you must configure baud rate, the number of data bits per character, the sense of the parity bit, and the number of stop bits. The baud rate is the transmission rate in bits/s. Data words are ordered so the least significant bit (LSB) is transmitted first and the most significant bit (MSB) is transmitted last. Furthermore, 1 is a negative voltage and 0 is a positive voltage. The parity bit is used for error checking. It can be set at transmission so the count of the number of ones in the data and parity bit is even or odd. If the receiver counts the ones and finds they are even when odd was set, an error is declared. The stop bits are negative voltages and can be 1, 1.5, or 2 bits long.

Instruments can also communicate over the universal serial bus (USB), which is usually available in PCs. It is "plug and play," meaning it automatically detects

TABLE 7.4   IEEE 488.2 Controller Protocols (from ni.com)

Description	Control Sequence	Compliance
Send ATN-true commands	SEND COMMAND	Mandatory
Set address to send data	SEND SETUP	Mandatory
Send ATN-false data	SEND DATA BYTES	Mandatory
Send a program message	SEND	Mandatory
Set address to receive data	RECEIVE SETUP	Mandatory
Receive ATN-false data	RECEIVE RESPONSE MESSAGE	Mandatory
Receive a response message	RECEIVE	Mandatory
Pulse IFC line	SEND IFC	Mandatory
Place devices in DCAS	DEVICE CLEAR	Mandatory
Place devices in local state	ENABLE LOCAL CONTROLS	Mandatory
Place devices in remote state	ENABLE REMOTE	Mandatory
Place devices in remote with local lockout state	SET RWLS	Mandatory
Place devices in local lockout state	SEND LLO	Mandatory
Read IEEE 488.1 status byte	READ STATUS BYTE	Mandatory
Send group execution trigger (GET) message	TRIGGER	Mandatory
Give control to another device	PASS CONTROL	Optional
Conduct a parallel poll	PERFORM PARALLEL POLL	Optional
Configure devices' parallel poll responses	PARALLEL POLL CONFIGURE	Optional
Disable devices' parallel poll capability	PARALLEL POLL UNCONFIGURE	Optional

TABLE 7.5   IEEE 488.2 Common Mandatory Commands (from ni.com)

Mnemonic	Group	Description
*IDN?	System Data	Identification query
*RST	Internal Operations	Reset
*TST?	Internal Operations	Self-test query
*OPC	Synchronization	Operation complete
*OPC?	Synchronization	Operation complete query
*WAI	Synchronization	Wait to complete
*CLS	Status and Event	Clear status
*ESE	Status and Event	Event status enable
*ESE?	Status and Event	Event status enable query
*ESR?	Status and Event	Event status register query
*SRE	Status and Event	Service request enable
*SRE?	Status and Event	Service request enable query
*STB?	Status and Event	Read status byte query

when a device has been added. Data rates up to 60 Mbytes/s can be achieved with USB 2.0 devices. The cables and connectors are not industrial grade, like the ones for GPIB, and the maximum cable length is 30 m.

Ethernet is another instrument communication bus that is common to PCs. It is usually built in to the motherboard, and it can transmit at rates of 1.25 Mbytes/s to 0.125 Gbytes/s. Ethernet facilitates remote control of instruments; however, network traffic can affect transfer, and there are security concerns. Determinism can be affected if the Ethernet is not dedicated only to instrument control. When two Ethernet devices start transmitting at the same time, their messages collide. If multiple nodes are waiting for a clear carrier, they all start transmitting at the same time and cause a collision. Collisions are a normal part of Ethernet messaging, and devices are equipped with Collision Detect. When a collision is detected, the nodes will stop transmitting and repeatedly try again after a random amount of time. This causes nondeterminism as the delay in transmission is random. Collisions can be minimized by sending small packets, reducing the size of the packets to only the information necessary, and limiting the number of nodes on a local network.

Another issue for measurement systems on the Internet is prioritizing the measurement task. For example, many users access the data from a measurement system connected to the Internet, and they can use all of the resources and the measurement systems computer resources, so it would not be able to continue taking measurements at the specified rate. You may need to isolate the measurement system from these inquiries by dedicating a computer to the measurement task and transmitting the data to a separate server. The server would then handle the inquiries from Internet users.

Instruments can communicate over FireWire, also. You may have to purchase a plug-in board for FireWire, and not many instruments support it. FireWire can transfer up to 0.4 Gbytes/s. Up to 16 devices can be connected within 4.5 m of each other for a total distance of 72 m. FireWire cables are not industrial grade, and data can be lost in transmission.

GPIB, serial, Ethernet, USB, and FireWire are stand-alone busses. In addition to stand-alone busses, there are modular busses, such as PCI, PCI Express, VXI, and PXI, where the bus is implemented in the instrument itself for higher performance. Refer to ni.com for more information on these busses.

Instrument Drivers

Instrument drivers reduce the effort to develop programs to communicate with instruments. Three common types of instrument drivers are LabVIEW plug and play, Interchangeable Virtual Instruments (IVI), and contributed. LabVIEW plug-and-play drivers maintain a common architecture and interface, so you can quickly connect to and communicate with an instrument with very little or no code development. Plug-and-play drivers are available in LabVIEW from the Instrument Driver Finder in Help>>Find Instrument Drivers, and from ni.com. When you install an instrument driver, an example program using the driver is added to the LabVIEW example finder.

IVI drivers are sophisticated, dynamic link library (DLL)-based drivers developed in LabWindows™/CVI™ that allow for simulation and instrument interchangeability.

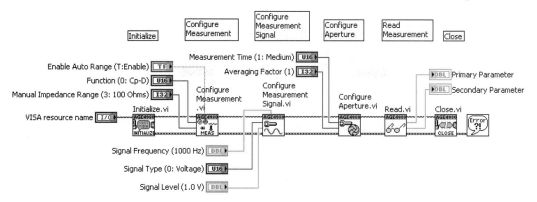

FIGURE 7.17 Agilent E4980 Read Measurement Example VI

Contributed instrument drivers may solve a specific application instead of being fully functional. Contributed instrument drivers are not supported by NI or other third parties. Developing and using IVI and contributed drivers are beyond the scope of this text.

LabVIEW plug-and-play instrument drivers usually include the following:

> ➤ Initialize VI—opens a resource that assigns memory that establishes communication with the instrument. Generally, you need to call the Initialize VI only once at the beginning of an application, as shown in Figure 7.17. The Initialize VI usually creates a resource reference that is passed to successive driver sub VIs in a LabVIEW application.
> ➤ Configuration VIs—a collection of software routines that configure the instrument to perform the operation you want. After you call these VIs, the instrument is ready to take measurements or to stimulate a system. There are three configuration VIs in Figure 7.17.
> ➤ Action VIs—order the instrument to carry out an action based on its current configuration, for example, the Read VI in Figure 7.17.
> ➤ Status VIs—obtain the current status of the instrument or the status of pending operations.
> ➤ Data VIs—transfer data to or from the instrument.
> ➤ Utility VIs—perform a variety of operations such as reset, self-test, revision, error query, error message, calibration, and storage and recall of setups.
> ➤ Close VI—terminates the software connection to the instrument and frees system resources. Generally, you need to call the Close VI only once at the end of an application or when you finish communication with an instrument. Each time the Initialize VI opens a resource, a Close VI should be used to close the resource to free memory assigned to the resource for other applications as shown in Figure 7.17.

Instrument drivers are usually sub VIs in a LabVIEW application, as shown in Figure 7.17.

LabVIEW instrument drivers usually use Virtual Instrument Software Architecture (VISA) functions to communicate with instruments. There are many types of instruments,

FIGURE 7.18
Agilent E4890
Fetch VI

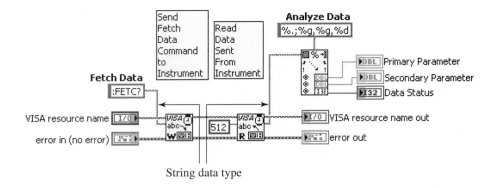

String data type

but you need to learn only one software architecture (VISA) to communicate with all of them. We can see the VISA functions used in the Agilent E4980 Read Measurement application by opening the Read VI and opening its sub VI, the Fetch VI, shown in Figure 7.18. The block diagram window of the Agilent E4980 Fetch VI uses the VISA Write function to request data from the instrument by sending the Fetch command. The VISA Read function returns an ASCII string of information that the Scan from String function converts to numerical values according to the format specification.

The front panel window of most instrument driver VIs includes a VISA resource name control and indicator. The block diagram icons for these objects are shown in Figure 7.18. As in other I/O operations, the resource is opened by the initialize VI and is passed between successive sub VIs. Instruments that are powered on and connected to your computer are visible in the VISA Resource name pull-down menu. You can create VISA names in MAX using the following format: Interface Type::Address:: INSTR; for example, GPIB::22::INSTR, GPIB board 0, primary address 22.

Sometimes instrument drivers are not available for the instrument you want to control. In that case, you can create your own instrument driver and contribute it to ni.com. Creating drivers is beyond the scope of this introductory text. See the instrument driver network on ni.com for more information. Alternatively, you can use the Instrument I/O Express VI.

Formatting and Building Command Strings

Instruments connected to GPIB, serial, USB, or Ethernet busses communicate with high-level ASCII character strings called messages. The instrument has a local processor that converts the messages to register bits. The messages usually follow the Standard Commands for Programmable Instruments (SCPI) standard that includes the commands in Table 7.5.

Some instruments require you write directly to the instrument control registers, and there are no standards for register-based instrument communication. So, you must use the instrument documentation to program each instrument. Register-level programming is beyond the scope of this text.

When you communicate with a message-based instrument, you must format and build the correct command strings for the instrument to perform the appropriate operation or return a response. Typically, a command string, or query, is a combination of text and numeric values. Some instruments require text-only command strings, requiring you to convert the numeric values to text and append them to the command string. Similarly, to use the data an instrument returns in LabVIEW, you must convert the data to a format that is useful for your application. Instrument drivers normally do the conversion for you, as shown in Figure 7.18.

Most instruments return data with headers and/or trailers. The header usually contains information such as the number of data points. Trailer information often contains units. You must remove the header and trailer information before you can display or analyze the returned data. The Instrument Control Parsing VI, shown in Figure 7.19, gives an example of plotting data from an instrument that transmitted a 2-byte binary format where 16 bits of data, or two ASCII characters, represent each data value.

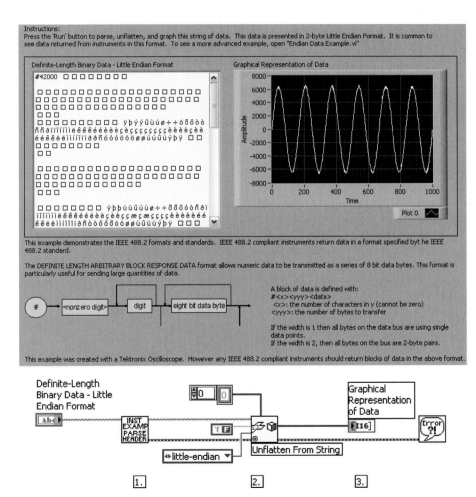

FIGURE 7.19
Instrument Control Parsing VI (from LabVIEW examples)

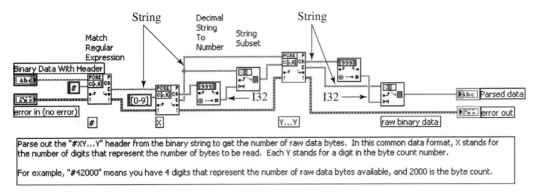

FIGURE 7.20 Instrument example Parse Header Sub VI Block Diagram

The returned data can be in ASCII, 1-byte binary, or 2-byte binary formats. If an instrument returns data in ASCII format, you can view the data as shown in Figure 7.18, and convert the string data to numeric data. If the data is in binary format, you must convert the binary string to a numeric array. If the instrument sends a binary string that contains 1024 1-byte binary encoded values, the waveform requires only 1024 bytes plus any header information. Using binary encoding, you need only 1 byte to represent each data value, assuming each value is an unsigned 8-bit integer.

If the instrument sent a signal of 1024 data points, the signal would consist of 2048 bytes because each value is a 2-byte signed integer. The instrument also sends a 5-byte header and a 2-byte trailer. So to view the data, the Instrument Control Parsing VI removes the 5-byte header using the sub VI shown in Figure 7.20. Then, it executes the Unflatten From String function to convert the waveform string to an array of 16-bit integers. Instruments can transmit data in big endian, little endian, or native order. In big-endian order, the most significant byte occupies the lowest memory address. In little-endian order, the least significant byte occupies the lowest memory address. In native order, the byte-ordering format of the host computer is used. You must know the order used by the instrument by studying its documentation and set the constant correctly in Unflatten From String function, as shown in Figure 7.19.

The information in the string control on the front panel window of the Instrument Control Parsing VI can be viewed in '\' codes or hex displays as shown in Figure 7.21.

Instrument I/O Assistant Express VI

If you don't have an instrument driver, you can use the Instrument I/O Assistant Express VI. Like the other Express VIs explained previously, it uses a dialog window to build code for you. When you place the Instrument I/O Assistant Express VI on the block diagram window, the dialog window opens, as shown in Figure 7.22. Then, select your instrument from the drop-down instrument list and add steps to communicate with it. The list of instruments is generated automatically by communicating with information from MAX.

FIGURE 7.21 '\' codes display

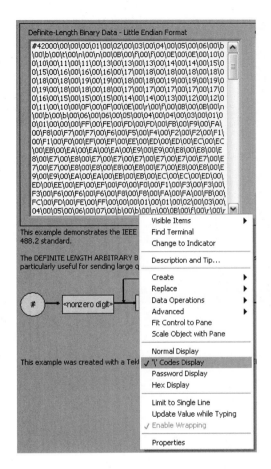

This example demonstrates the IEEE
488.2 standard.

The DEFINITE LENGTH ARBITRARY B
particularly useful for sending large q

This example was created with a Tek

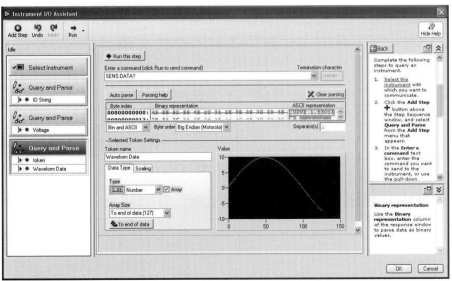

FIGURE 7.22
Interactive Instrument
I/O Assistant Express
VI dialog window

FIGURE 7.23 Instrument I/O
Assistant in a block diagram
window

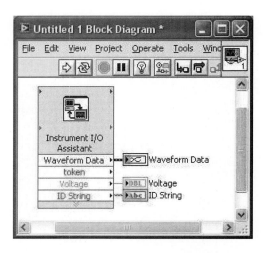

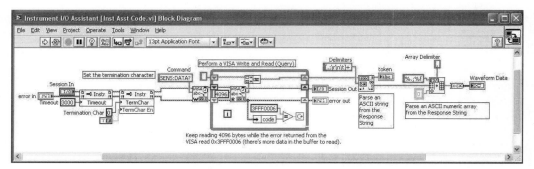

FIGURE 7.24 Code developed by the Instrument I/O Assistant

You can communicate with the instrument in a series of steps that you define. Steps can include write, query, and parse commands. The result is automatically displayed in the interactive window. After you click OK, the Instrument I/O Assistant becomes an icon on a block diagram window, as shown in Figure 7.23.

You can convert your Instrument I/O Assistant Express VIs to standard LabVIEW VIs to view and/or use the underlying code and further understand the parsing and formatting, as shown in Figure 7.24.

CLIENT/SERVER COMMUNICATIONS BETWEEN COMPUTERS

LabVIEW supports distributed applications where the I/O operations occur over a network. Typical tasks include the following:

> Share live data with other VIs running on a network using shared variables.
> Publish front panel images and VI documentation on the Web.

Server program

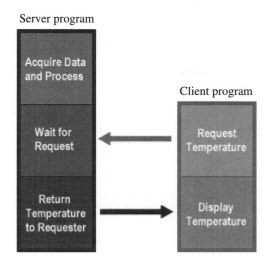

Client program

FIGURE 7.25 Client-server
application example (after
Understanding Client-Server
Applications—Part I at ni.com)

➢ E-mail data from VIs.
➢ Build VIs that communicate with other applications and VIs through low-level
   protocols, such as TCP, UDP, Apple events, and PPC Toolbox.

A common application architecture used for these communications establishes
one computer as the client that requests an action or service from another computer,
the server. We previously explained that modular programming separates large applications into small modules called subprograms that decompose large problems to
facilitate development, testing, and maintenance. In a client-server application, a
module does not have to be part of the same program or even run on the same computer. Each modular function can run on a different device.

Client programs request service from a server by sending it a message. Server
programs process client requests by performing the tasks requested by clients. Servers are generally passive as they wait for a client request. During these waiting periods, servers can perform other tasks or perform maintenance. Unlike the client, the
server must continually run because clients can request service at any time. Clients,
on the other hand, need to run only when they require service. Many server applications allow for multiple clients to request service. This type of application can be
implemented with a client-server architecture, shown in Figure 7.25.

In the simplest application, only one remote client requests information from a
server. In this case, the server waits for a request and returns the information to the
requesting client. If multiple clients are serviced, the server continually monitors for
service requests. See Understanding Client-Server Applications—Part I at www.ni.com
for additional information.

COMMUNICATING OVER THE INTERNET WITH TCP/IP

Networks can be extremely large and connect a wide variety of computers with varying
hardware and software. Consequently, standard protocols specify how data will be
formatted, flow will be controlled, information exchanged, and errors handled.

Internetworking Protocol (IP) prescribes how the many varieties of hardware and software intercommunicate, and Transmission Control Protocol (TCP) prescribes how data are routed and communicated error free. The two protocols are commonly combined into TCP/IP.

You can use TCP/IP to communicate over single networks or interconnected networks. The individual networks can be separated by large geographical distances. TCP/IP routes data from one network or Internet-connected computer to another. Because TCP/IP is available on most computers, it can transfer information among diverse systems.

IP performs the low-level service of moving data between computers. IP packages data into components called datagrams. A datagram contains the data and a header that indicates the source and destination addresses. IP determines the correct path for the datagram to take across the network or Internet and sends the data to the specified destination.

IP cannot guarantee delivery. In fact, IP might deliver a single datagram more than once if the datagram is duplicated in transmission. Programs rarely use IP but use TCP or UDP (user Datagram Protocol) instead. TCP ensures reliable transmission across networks, delivering data in sequence without errors, loss, or duplication. TCP retransmits the datagram until it receives an acknowledgment.

TCP is a connection-based protocol, which means that sites must establish a connection before transferring data. TCP permits multiple, simultaneous connections. Initiate a connection by waiting for an incoming connection or by actively seeking a connection with a specified address and a port at that address. Different ports at a given address identify different services at that address. We implement the TCP protocol in the following steps:

1. A server that listens on a selected port and creates a connection when the client initiates a request.
2. Client applications that connect to the selected port, decode the data, and display it.

A port is the location where you send data.
Create a TCP server like the one shown in Figure 7.26:

1. Use the TCP Listen VI or the TCP Create Listener and TCP Wait on Listener functions to wait for a connection. Specify the port and optionally a service name, which is the same one that the client attempts to access.
2. If a connection is established, use the TCP Write function to send a message to the client. The data must be in a form that the client can accept.
3. Read a message from the client using the TCP Read function.
4. Close the connection with the TCP Close Connection function.

Create a TCP client like the one shown in Figure 7.27:

1. Use the TCP Open Connection function to open a connection to a server. You must specify the remote port or service name for the server whose name identifies a communication channel on the computer that the server listens to for communication requests. If you specify a service name, LabVIEW queries the NI Service Locator for the port number that the server registered.

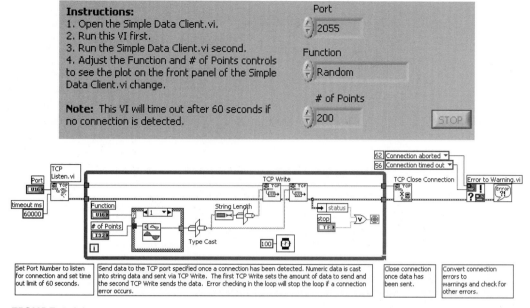

FIGURE 7.26 Simple Data Server Example VI (LabVIEW Examples)

2. If you want to establish a connection to a remote computer, you must specify the Internet address of the server. If you do not specify an address, LabVIEW establishes a connection to the local computer.
3. Use the TCP Read function to read a message from the server. You must specify the number of bytes you want to read.
4. Use the TCP Write function to send a message to a server.
5. Use the TCP Close Connection function to close the connection to the server.

Use the TCP Open Connection function to actively establish a connection with a specific address and port. If the connection is successful, the function returns a network connection refnum that uniquely identifies that connection. Use this connection refnum to refer to the connection in subsequent VI calls.

You can use the following techniques to wait for an incoming connection:

➢ Use the TCP Listen VI to create a listener and wait for an accepted TCP connection at a specified port. If the connection is successful, the VI returns a connection refnum, the address, and the port of the remote TCP client.
➢ Use the TCP Create Listener function to create a listener, and use the TCP Wait on Listener function to listen for and accept new connections. The TCP Wait On Listener function returns the same listener ID you wired to the function and the connection refnum for a connection. When you finish waiting for new connections, use the TCP Close Connection function to close a listener. You cannot read from or write to a listener.

The advantage of using the second technique is that you can use the TCP Close Connection function to cancel a listen operation, which is useful when you

FIGURE 7.27
Simple Data Client
Example VI (LabVIEW
Examples)

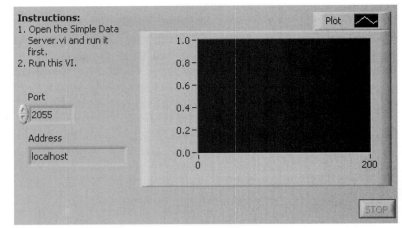

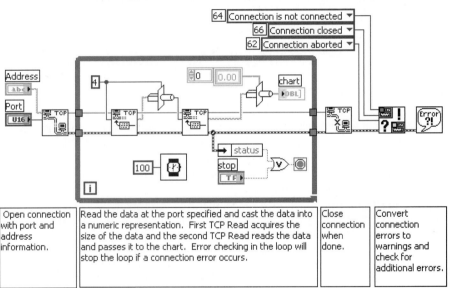

want to listen for a connection without using a timeout, but you want to cancel the listen when another condition becomes true. You can close the listen VI at any time.

When you establish a connection, use the TCP Read function and the TCP Write function to read and write data to the remote application.

Use the TCP Close Connection function to close the connection to the remote application. If unread data remains and the connection closes, you might lose data. Use a higher-level protocol for your computer to determine when to close the connection. After a connection is closed, you cannot read from it or write to it again.

When you design a network application, consider carefully what should happen if something fails. For example, if the server crashes, determine how each client VI handles it. One solution is to make sure that each client VI has a timeout. If something fails to produce results after a certain amount of time, the client continues execution. In continuing, the client can try to reestablish the connection or report the error. If necessary, the client VI can shut down the application.

COMMUNICATING OVER THE INTERNET WITH UDP

LabVIEW also supports User Datagram Protocol (UDP). UDP provides simple, low-level communication among processes on computers. Use UDP in applications in which reliability is not critical. For example, an application might transmit informative data to a destination frequently enough that a few lost segments of data are not problematic. Processes communicate by sending datagrams to a destination computer or port. IP handles the computer-to-computer delivery. After the datagram reaches the destination computer, UDP moves the datagram to its destination port. If the destination port is not open, UDP discards the datagram. UDP shares the same delivery problems of IP. You can use the UDP functions to communicate to a single client (single-cast) or to all computers on the subnet through a broadcast. Use multicasting to communicate between a single sender and multiple clients on a network without requiring the sender to maintain a list of clients or send multiple copies of the data to each client. To receive data broadcast by a multicast sender, all clients must join a multicast group. The sender does not have to join a group to send data. The sender specifies a multicast IP address, which defines a multicast group. Multicast IP addresses are in the 224.0.0.0 to 239.255.255.255 range. When a client wants to join a multicast group, it subscribes to the multicast IP address of the group. After subscribing to a multicast group, the client receives data sent to the multicast IP address. Refer to the following VIs for examples of using UDP multicasting:

> UDP Multicast Receiver VI: labview\examples\comm\UDP.llb

> UDP Multicast Sender VI: labview\examples\comm\UDP.llb

TCP is the best protocol to use if you want reliable data transmission. UDP is a connectionless protocol with higher performance, but it does not ensure reliable data transmission.

SUMMARY

> Computers are useful when they interact with the world through input and output (I/O) units or resources. This chapter describes methods of interacting with resources such as files, sensors, instruments, and other computers over networks.

> The basic paradigm used throughout the chapter for data exchange with many resources is to open and/or configure the resource, read or write to it, clear or close it, and handle errors.

> We use the American Standard Code for Information Interchange (ASCII) format, or strings, to format data to be transferred to several resources such as some files and instruments.

> We use a format code to specify how to convert numeric data to or from a string.

➤ LabVIEW can also read and write files in binary and datalog formats as well as ASCII format.

➤ We wire refnum and error cluster data from open file to read or write file and finally to close file VIs to communicate information about the file, handle errors, and enforce sequential data flow.

➤ LabVIEW provides high-level VIs such as the Write to Spreadsheet VI and the Write Lab-VIEW Measurement File Express VI to simplify data exchange with files.

➤ We can use disk streaming for more efficient data exchange where we use lower-level open, read or write, and close functions. Consequently, we open the file once, write to it many times in a repetition structure, and close it once.

➤ Error checking in a program provides information on why and where errors occur. We can use this information to handle errors in appropriate ways.

➤ Communicating with hardware located externally to a computer can be very complicated because of the wide range of hardware devices that can be connected and the wide range of ways they can be connected. This chapter presented a simple introduction to writing programs in Lab-VIEW that support some simple applications.

➤ We can use DAQmx VIs with the same concepts as disk streaming. That is, we can open a DAQ device, start acquisition, read the data from memory repetitively, clear the process, and handle errors.

➤ Information exchange between computers over a network is a common task for engineers and scientists. A common application architecture used for these communications establishes one computer as the client that requests an action or service from another computer, the server.

➤ TCP and UDP functions are available in Lab-VIEW for developing network communications programs.

➤ In addition to communicating between computers, we can communicate with external instruments such as oscilloscopes.

➤ Instruments commonly communicate with computers over GPIB, or Serial Bus, USB and Ethernet.

➤ Instrument drivers can be implemented in Virtual Instrument Software Architecture (VISA) in LabVIEW using the same I/O paradigm: open communications, read or write, clear communications, and handle errors.

➤ The Instrument Assistant Express VI can be used when instrument driver software is not available.

EXERCISES

1. Update your course notebook with information from this chapter, including new data types and shortcut keys.

2. Read the chapter and modify the Temperature Measurement State Machine to save voltage data to file as each point is acquired with disk streaming. Test it.

3. Add a Save State to the Climate Simulation state machine that saves the data after 24 data points are acquired. Use the Write to Text File function to save the data.

4. Modify the Climate Simulation state machine so the statistics are added to a cluster that also includes the Temperature and Humidity Values 2-D Array. Wire the cluster to a shift register so it is available to all states.

5. Save the statistical information along with the Temperature and Humidity 2-D Array in the Climate Simulation State Machine to the same file. Modify the VI to save to two different files.

6. Experiment with error handling. Create some intentional File errors using the Temperature State Machine and observe and report on your experiment.

7. Modify the Temperature Units Conversion subprogram to have error input and output terminals. Enclose the code in a Case Structure so it executes only when there is no error input.

8. Study the Write to Text File.vi example program. Explain how the tab character constant is inserted and

explain why it is necessary. Explain what is written during the first iteration of the loop. What is written in the second iteration? Does the second iteration write over or append to the first iteration data?

_____

_____

_____

9. Add column headers to the file for the 2-D array and the statistical values in the Climate Simulation State Machine. Consider adding an initialize state to the State Machine for this text write operation.

10. Write a VI that will write exactly 100 temperature measurements to a spreadsheet file after all of the measurements are placed in an array. Run it while placing your finger over the sensor to change the temperature. Open the file in notepad. Do not use a spreadsheet program.

11. Write a VI that will read the file produced in Exercise 10 into an array. Make sure the file is a .txt type (ASCII or text file) and not a spreadsheet file. Average chunks of 20 values and write the averages to a second file and also to an array indicator on the front panel window.

12. Write a VI that will write exactly 100 temperature measurements to a Measurement File after all of the measurements are placed in an array. Use the Write to Measurement File Express VI. Run it while placing your finger over the sensor to change the temperature. Open the file in a text editing and in a spreadsheet program. How does this file differ from the ones in the above exercises?

_____

_____

13. Write a VI that will read the Measurement File produced in Exercise 12 into an array. Use the Read from Measurement File Express VI. Average chunks of 20 values and write the averages to a second file and also to an array indicator on the front panel window of the VI.

14. Develop a state machine application of your own that writes to or reads from a file.

15. Display the Temperature and Humidity Array in a Table Indicator. Consider using the Number to Decimal String Function.

16. Open the Basic Spectral Measurements VI from the LabVIEW examples. Add the Report Express VI from the Report Generation Subpalette. Create a report that displays the graphs of both signals and includes the report title, author, company name, and comments about the report. Give the user the option to enter additional comments. Save the report as an HTML file in c:\temp. Do not send to a printer. Run the VI and view the HTML document created by the VI.

17. Download the price of Intel stock over the period January 5, 2004 to today from yahoo.com and save it to a file on c:\temp. Create a LabVIEW program that will read the file and graph the price vs. time. Calculate an N-day running average (where the user inputs N) and graph it. Identify the potential sell and buy points by listing below the dates when the price fell below or rose above the running average.

_____

_____

_____

_____

_____

_____

High-Frequency DAQ

INTRODUCTION

Many DAQ applications (acceleration and noise, for example) require that we sample at high rates. High-frequency sampling requires efficient code that can execute fast. It also creates extremely large data files. Therefore, we must be careful when selecting the sampling rate and in structuring our applications. This chapter applies the fundamentals of high-frequency sampling in a music application. The next chapter summarizes the material by building a versatile and efficient DAQ state machine application.

OUTLINE

DAQMX CONTINUOUS ACQUIRE AND GRAPH WITH INTERNAL CLOCK

So far we have acquired signals that didn't change rapidly. However, many signals, such as sound and vibration, have much higher frequencies and must be sampled faster. We will begin by reviewing the LabVIEW example: DAQmx Cont Acq&Graph Int-Clk, shown in Figure 8.1.

This example uses the DAQmx VIs introduced in Chapter 7 in a way that facilitates high-speed sampling. The DAQmx Timing VI (identified as 2 in the figure) configures the number of samples to acquire or generate and creates a buffer when needed. The DAQmx Start Task VI configures the device to begin sampling as soon as the DAQmx Read VI executes. If you do not use this VI, a measurement task starts automatically when the DAQmx Read VI runs. If you do not use the DAQmx Start Task VI when you use the DAQmx Read VI multiple times, such as in a loop, the task starts and stops repeatedly, which reduces the performance of the application. Note that when this VI is used, the DAQmx Stop Task VI must be placed after the while loop, as shown in Figure 8.1. Our application becomes hardware timed with these VIs, whereas the previous programs we wrote were timed with software, particularly the Delay in the While Loop. Hardware timing uses the internal clock on the DAQ board, hence the Int-Clk in the program's name. The NI PCI-6024E plug-in DAQ boards can sample up to 200 kS/s per channel.

This program configures the acquisition rate before the While Loop begins to iterate. The DAQmx Read VI in the While Loop reads data into computer memory. So, if the While Loop runs too slowly, a lot of data will pile up in memory waiting to be read, graphed, analyzed, saved to file, and so on. The memory used for this purpose is called a buffer. If the buffer is too small, it will fill up. This application uses a circular buffer that will be written over starting at the beginning if the buffer fills up. Therefore, we could lose some data. Consequently, we must carefully design the application, so the While Loop executes quickly and the buffer is large enough to avoid losing data.

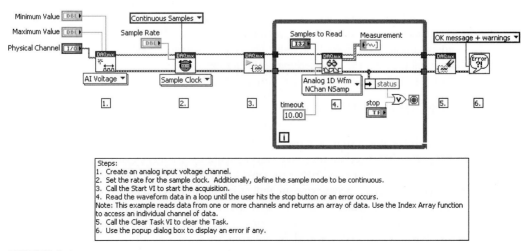

Steps:
1. Create an analog input voltage channel.
2. Set the rate for the sample clock. Additionally, define the sample mode to be continuous.
3. Call the Start VI to start the acquisition.
4. Read the waveform data in a loop until the user hits the stop button or an error occurs.
Note: This example reads data from one or more channels and returns an array of data. Use the Index Array function to access an individual channel of data.
5. Call the Clear Task VI to clear the Task.
6. Use the popup dialog box to display an error if any.

FIGURE 8.1 DAQmx Continuous Acquire and Graph Internal Clock Program block diagram window (LabVIEW Examples)

SAMPLING RATE

As previously mentioned, sampling at 200 kS/s fills up memory quickly. It also provides a large amount of data to graph, analyze, and store to files. Therefore, we want to minimize the sampling rate, but make sure we sample fast enough to truthfully represent the signal. We can use the Nyquist theorem, which states that the highest frequency we can accurately represent is equal to half the sampling rate.

$$f_N = \frac{f_s}{2},$$

where f_N is the Nyquist frequency and f_s is the sampling frequency. This means that if we want to measure the frequency of a 100 Hz signal, we need a sampling rate of at least 200 S/s.

The Nyquist theorem is most applicable to signals represented with only one frequency, like electrical power transmission signals. However, applications, like sound measurement, will generate several frequencies in one signal. Consequently, we need to be concerned with the highest frequency of interest, including the transition band. Furthermore, we might not be able to predict the frequencies in the signal. In practice, we might use higher sampling rates to begin with while we characterize the signal and conduct longer term tests later with lower frequencies.

If we take this approach, we should sample a minimum of three cycles, but 10 or more cycles, is better. For example, we might design an application to collect at least 15 samples, at a sampling rate of 500 S/s to measure the frequency of a 100 Hz signal. Because we are sampling about five times faster than the signal frequency, we sample about five points per cycle of the signal. Because we need data from at least three cycles, 5 points · 3 cycles = 15 points, at least. The number of points we collect determines the number of frequency bins that the data falls into. The size of each bin is the sampling rate divided by the number of points we collect. For example, if we sample at 500 S/s and collect 100 points, we will have bins at 5 Hz intervals.

If we do not sample fast enough, our result is aliased. An aliased signal provides a poor representation of the analog signal. Aliasing causes a false lower-frequency component to appear in the sampled data of a signal. Figure 8.2 shows examples of undersampled, or aliased, signals. The undersampled signals appear to have lower frequencies than the actual signal. We can use the preceding information to set the Sample Rate Control in Figure 8.1 to avoid aliasing.

FREQUENCY DOMAIN

The DAQ acquires signals in the time domain, that is, the amplitude of a signal in volts at a particular instant in time. We can calculate the frequency of the signal waveform by measuring the period between peaks or axis crossings and taking the inverse of the period. Many waveforms in industrial environments have noise and multiple frequencies that make it difficult to determine the various frequencies represented in the waveform and the intensity (or portion of the total power) at each frequency.

We can build the Signal Simulation VI shown in Figure 8.3, which allows us to simulate and display frequencies in signals. When we place the Spectral

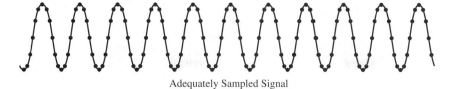

Adequately Sampled Signal

FIGURE 8.2
Undersampled, or
aliased, signals
(LabVIEW Help
>Aliasing)

Aliased Signal Due to Undersampling

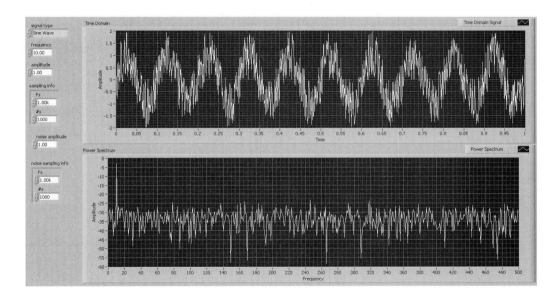

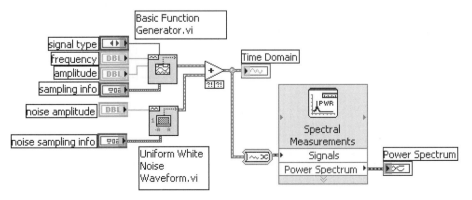

FIGURE 8.3 Signal Simulation VI

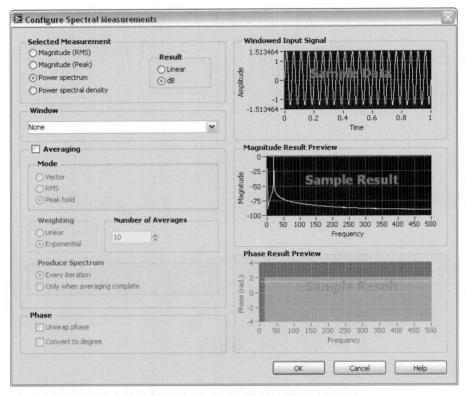

FIGURE 8.4 Configure Spectral Measurements window

FIGURE 8.5
Samples of a
constant 1 V signal

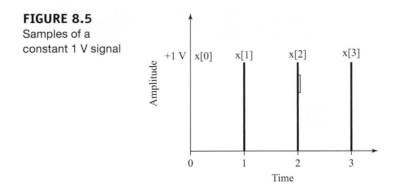

Measurements Express VI on the Block Diagram, the configuration window appears, as shown in Figure 8.4. Configure it as shown. The Signal Simulation VI uses the Spectral Measurement Express VI to represent a Time Domain Signal in the

Frequency Domain. There are numerous mathematical approaches using the process to transform time domain data to the frequency domain. One of the most common is the Fourier transform.

The Fourier transform of a signal x(t) is expressed as

$$X(f) = F(x\{t\}) = \int_{-\infty}^{\infty} x(t)e^{-j2\pi ft}\, dt$$

Fourier's theorem states that any waveform in the time domain can be represented by the weighted sum of sines and cosines. The same waveform then can be represented in the frequency domain as a pair of amplitude and phase values at each component frequency.

You can generate any waveform by adding sine waves, each with a particular amplitude and phase. After transformation, a sequence of n data points can be represented exactly by the summation of n sampled sine waves. The sine waves have frequencies N/T, where $N = 0, 1, \ldots, N - 1$, and T is the sampling period.

The Spectral Measurement Express VI uses the discrete form (the DFT of Discrete Fourier Transform), which is a summation from 0 to $N - 1$, rather than integration:

$$X_k = \sum_{i=0}^{N-1} x_i e^{-j2\pi ik/N} \qquad (1)$$

where $k = 0, 1, \ldots, N - 1$, and $X[0]$ is the DC or average value of the signal.

LabVIEW online help for the DFT presents an example that calculates the DFT for a 1 V DC signal sampled four times, as shown in Figure 8.5.

Each of the samples has a value $+1$, giving the time sequence $x[0] = x[1] = x[3] = x[4] = 1$. Using Equation (1) to calculate the DFT of this sequence and making use of Euler's identity, $\exp(-i\theta) = \cos(\theta) = j\sin(\theta)$ you get

$$X[0] = \sum_{i=0}^{N-1} x_i e^{-j2\pi i0/N} = x[0] + x[1] + x[2] + x[3] = 4$$

$$X[1] = x[0] + x[1]\left(\cos\left(\tfrac{\pi}{2}\right) - j\sin\left(\tfrac{\pi}{2}\right)\right) + x[2](\cos(\pi) - j\sin(\pi)) + $$
$$x[3]\left(\cos\left(\tfrac{3\pi}{2}\right) - j\sin\left(\tfrac{3\pi}{2}\right)\right) = (1 - j - 1 + j) = 0$$

$$X[2] = x[0] + x[1](\cos(\pi) - j\sin(\pi)) + x[2](\cos(2\pi) - j\sin(2\pi)) + $$
$$x[3](\cos(3\pi) - j\sin(3\pi) = (1 - 1 + 1 - 1) = 0$$

$$X[3] = x[0] + x[1]\left(\cos\left(\tfrac{3\pi}{2}\right) - j\sin\left(\tfrac{3\pi}{2}\right)\right) + x[2](\cos(3\pi) - j\sin(3\pi))$$
$$x[3]\left(\cos\left(\tfrac{9\pi}{2}\right) - j\sin\left(\tfrac{9\pi}{2}\right)\right) = (1 - j - 1 - j) = 0$$

As expected, all values except $X[0]$ are zero. $X[0]$ depends on N, which is 4 in this case, so $X[0] = 4$. Thus, we divide $X[0]$ by N to obtain the correct magnitude of the frequency component.

POWER SPECTRUM AMPLITUDE

The configuration window gives us the option of displaying amplitude on either a linear scale or a decibel (dB) scale. Displaying amplitude or power spectra on a decibel scale allows us to view wide dynamic ranges and to see small signal components in the presence of large ones. If we wanted to display a range of 0.1 V to 100 V on a graph 10 cm high, the graph would display 10 V/cm on a linear scale, and the 0.1 V amplitude would be only 0.1 mm high, barely visible. But a logarithmic scale in decibels allows us to see the 0.1 V amplitude much better.

The decibel is a unit of ratio. The decibel scale is a transformation of the linear scale into a logarithmic scale. For a ratio of powers, the decibel is defined as

$$dB = 10\log_{10}\frac{P}{P_r},$$

where P is the measured power, P_r is the reference power, and P/P_r is the power ratio.

The decibel in terms of voltage ratio is

$$dB = 20\log_{10}\frac{A}{A_r},$$

where A is the measured amplitude, A_r is the reference amplitude, and A/A_r is the voltage ratio.

MULTIPLE LOOPS

We introduced the multiple loop architecture in Figure 5.6. A poorly coded parallel-loop architecture is shown in Figure 8.6. At first glance, we see that this program intends to acquire data in the loop on the top and display it in the loop on the bottom, stopping both loops with one control. This should allow the upper loop to run unimpeded by the slower user interface graph updating task. However, when we run the program, it doesn't work as intended because whoever coded it forgot about the LabVIEW data-flow principle. When we run it, the lower loop does not

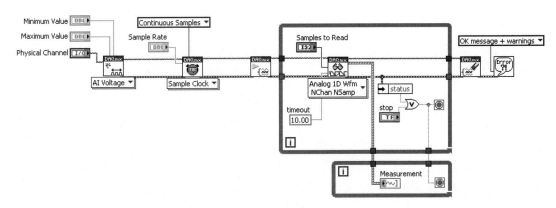

FIGURE 8.6 Incorrect Parallel Loops program

run in parallel (simultaneously) with the upper loop. This occurs because the lower loop won't start until it receives values in its input tunnels. However, the upper loop doesn't send out these values until it stops. Furthermore, since the first loop may have terminated with a true value from the stop control or the error cluster, the lower loop might run only once because it received a true stop value from the upper loop. Code the program and run it with Execution Highlighting to verify the data flow.

VARIABLES

Figure 8.7 shows an attempt to solve the problem of exchanging data between loops with local variables. Local variables pass information between locations in a program that cannot connect with a wire. They are a location in computer memory that can be written to and read from. The program can use local variables to

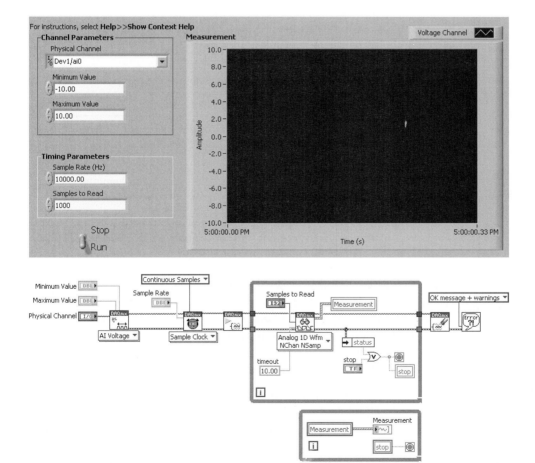

FIGURE 8.7 Local Variable program

access front panel objects from more than one location in a single VI. Because they do not require wires, they can solve our problem in the Parallel Loops program.

We can use global variables instead of local variables to access and pass data between VIs instead of between loops in a VI. Refer to LabVIEW Help for information on using global variables.

Local and global variables cause problems with race conditions and synchronization, so don't use them unless it is absolutely necessary. Read the following to learn more about the problems with variables and about alternative techniques such as queues.

To create a local variable, right-click an existing front panel object or block diagram terminal and select Create»Local Variable from the shortcut menu. You can also select a local variable from the Functions Palette and place it on the block diagram. The local variable node is not yet associated with a control or indicator. To associate a local variable with a control or indicator, right-click the local variable node and select Select Item from the shortcut menu. The expanded shortcut menu lists all front panel objects that have owned labels.

LabVIEW uses owned labels to associate local variables with front panel objects, so label the front panel controls and indicators with descriptive owned labels.

MECHANICAL ACTION OF BOOLEANS

Note that the traditional Stop button was replaced by a Boolean toggle control in Figure 8.7. The typical Stop button uses a mechanical latching action that cannot be used for objects with a local variable because the first local variable to read a Boolean control with latch action would reset its value to the default. Boolean controls have six different behaviors (mechanical actions):

1. Switch When Pressed—Changes the control value each time you click it with the Operating tool, similar to a light switch. The action is not affected by how often the VI reads the control.
2. Switch When Released—Changes the control value only after you release the mouse button during a mouse click within the graphical boundary of the control. The action is not affected by how often the VI reads the control.
3. Switch Until Released—Changes the control value when you click it and retains the new value until you release the mouse button. At this time, the control reverts to its original value, similar to the operation of a door buzzer. The action is not affected by how often the VI reads the control. You cannot select this action for a radio buttons control.
4. Latch When Pressed—Changes the control value when you click it and retains the new value until the VI reads it once. At this point, the control reverts to its default value even if you keep pressing the mouse button. This action is similar to a circuit breaker and is useful for stopping a While Loop or for getting the VI to perform an action only once each time you set the control. You cannot select this action for a radio buttons control.

5. Latch When Released—Changes the control value only after you release the mouse button within the graphical boundary of the control. When the VI reads it once, the control reverts to the old value. This action works in the same manner as dialog box buttons and system buttons. You cannot select this action for a radio buttons control.

6. Latch Until Released—Changes the control value when you click it and retains the value until the VI reads it once or you release the mouse button, depending on which one occurs last. You cannot select this action for a radio buttons control.

Note that the While Loop Conditional Terminal was changed from Stop if True to Continue if True in the preceding programs, so the program would run when the toggle was up, or True.

PROBLEMS WITH VARIABLES

Because local variables don't use wires, they break LabVIEW's data-flow paradigm and are not an optimal solution. Writing to a local variable is similar to passing data to any other terminal. However, with a local variable, the program can write to it even if it is a control or read from it even if it is an indicator. The local variable used to stop the parallel loops in Figure 8.7 is a good use of a local variable. However, the other local variable can exhibit problems with synchronization. To demonstrate the synchronization problem, we will add a graph to the first loop, as shown in Figure 8.8. The synchronization problem occurs when the loops run at different speeds. If the delay for the first loop is 100 ms and that for the second is 200 ms, the

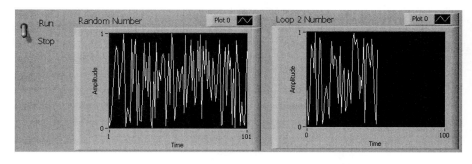

FIGURE 8.8
Synchronization problem with variables

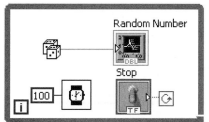
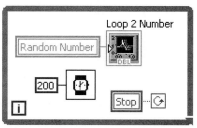

graph will skip the display of every other point. This occurs because the first loop writes over the value in memory before the second loop reads it. If the first loop delay is 200 ms and the second is 100 ms, the graph will duplicate the display of every point. This occurs because the second loop reads the value twice before it is written by the first loop.

In other instances, local and global variables can exhibit a problem called a race condition. A race condition occurs when two or more pieces of code that execute in parallel change the value of the same shared resource, typically a local or global variable. Race conditions are very difficult to debug because the program may work fine most of the time but will give an erroneous result sporadically.

PRODUCER CONSUMER DESIGN PATTERN AND QUEUES

Because we can't wire between multiple loops and we would like to avoid synchronization and race problems with local and global variables, we need another technique to share data between loops. The Producer Consumer Design Pattern template (File>>New>>Frameworks>>Design Patterns), shown in Figure 8.9, is one solution. The Producer Consumer Design Pattern uses queues to exchange data. Queues "stack" data into memory in order. Data are retrieved in a first in, first out (FIFO) order. Consequently, it doesn't exhibit the synchronization and race condition problems of local and global variables. Use the Queue Operations functions (Functions>>Programming>>Synchronization) to store data in a queue,

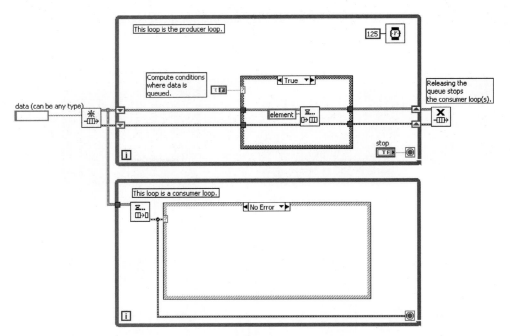

FIGURE 8.9 Producer Consumer Design Pattern

which the program can later retrieve as individual elements or as an array of all the data elements.

SIGNAL CONDITIONING FOR SIMPLE SOUND MEASUREMENT

Waveforms that are very low in amplitude, or accompanied with noise, may need to be amplified or preprocessed with analog circuitry before entering the DAQ boards for digital sampling. The most common signal conditioning processes involve filtering, removing DC components from signals, impedance matching, and amplifying. High-voltage signals require attenuation so that they will not exceed the input ratings of the DAQ circuitry.

We will build a microphone circuit with a protoboard and various electrical components to experiment with some signal conditioning methods for simple sound measurement.

The protoboard (also known as a solderless breadboard) is a device that facilitates experimental circuit construction. Figure 8.10 shows the protoboard as it appears with no components attached. Protoboards are available on the web and at local electronic shops. We used a Jameco 20601. The board contains holes that can receive a wire or a component lead and hold it in place for connection to other wires or components. As shown in Figure 8.11, the lines indicate how the top and bottom two rows of holes and many small vertical columns of five holes each are connected together underneath the board and out of view. The long horizontal rows are called

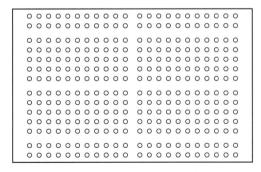

FIGURE 8.10
Top view of protoboard showing holes for wires and components

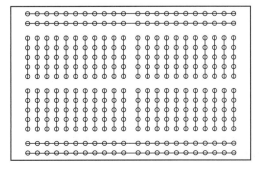

FIGURE 8.11
Bottom view of protoboard showing wiring beneath the surface that connects the holes

FIGURE 8.12

Microphone wires

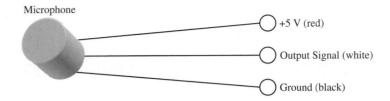

bus lines and are usually reserved for wires and components that need to be connected to common nodes that may have many items connected to them, including external connections, such as a ground line, a signal input or output line, and/or a power supply line. The smaller vertical columns are generally used as nodes for connecting components and wires internally in the circuit.

To construct a basic microphone circuit, begin by placing the three leads of an electret microphone into the protoboard, as shown in Figure 8.12, Figure 8.13, and Figure 8.14. Electret microphones are available on the web and at local electronics stores. We used a Radio Shack Part Number 270-092.

1. Place the red wire (excitation voltage, +5 V) into one of the protoboard busses (top rows) that will connect to the external +5 V source on the BNC-2120.
2. Next, place the black wire (ground) into one of the bottom rows, which will connect to the external ground on the BNC-2120 and other ground wires.
3. Finally, place the microphone's output signal (white) wire into one of the internal holes on a convenient column of five holes of the protoboard.

FIGURE 8.13

Microphone circuit wiring on the protoboard

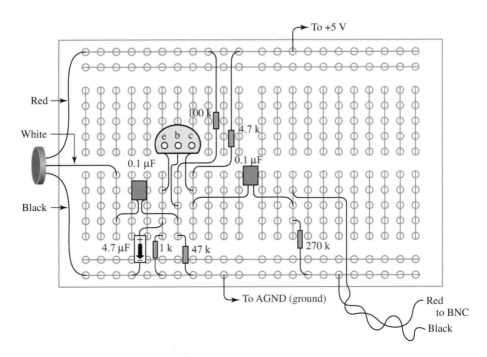

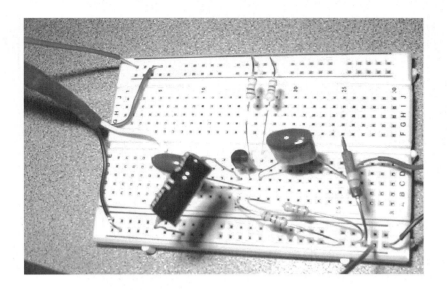

FIGURE 8.14
Photograph of the
microphone circuit

The next step in the process is to wire the BNC-2120 to power and acquire data from the microphone.

1. Connect a wire from the BNC-2120's 5 V output to the same bus as the microphone's red wire. Refer to the BNC-2120 information in Chapter 3.
2. Similarly do this for the ground by connecting the BNC-2120's analog ground to the bus containing the microphone's ground.
3. Finally, connect a wire from the column containing the microphone's output to the BNC-2120's analog input channel 0 (AI0) using a BNC-2120 through a twisted pair wire to a banana jack-to-BNC adapter. BNC adapters are available on the web and at local electronic shops. We used a Jameco 125233. Additionally, the BNC input is a differential measurement, so you will need to ground one part of this measurement.
4. Do not connect any of the other components shown in Figure 8.13 and Figure 8.14 at this point. We will build the circuit a section at a time and measure the signal sequentially throughout this discussion to show the effects of each portion of the circuit. Building the hardware circuit in modules and testing each module is similar to modular programming with sub VIs. If you divide any problem, whether it be hardware or software, into modules, then build and test each module, the final system will be composed of tested modules and will be more likely to function correctly than if we build the entire system before doing any testing.

We will modify the Signal Simulation VI shown in Figure 8.3 to acquire, analyze, and present the sound data. Replace the Basic Function Generator and the Uniform White Noise Express VIs with the DAQ Assistant to create the VI shown in Figure 8.15.

Configure the DAQ Assistant with the appropriate channel for an analog voltage input. Read in 1000 samples at a time and sample at a rate of 50,000 Hz, as shown in Figure 8.16. Enclose the code in a While Loop for continuous measurement. To view

FIGURE 8.15
Audio Measurement
program

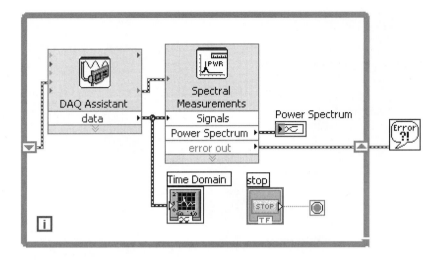

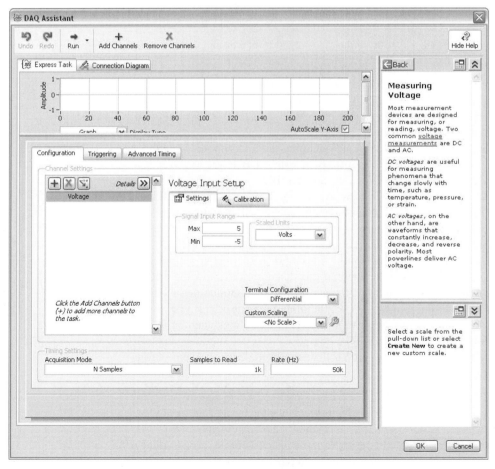

FIGURE 8.16 Audio Measurement program DAQ Assistant configuration

the appropriate frequencies, change the x-scale of your plot to show frequencies below 10,000 Hz.

When you run the program and attempt to measure common noises, you will notice that there are two problems with the signal fed directly from the microphone. First, the signal has a DC component that makes it difficult to see any very small change in the signal level. Whistling into the microphone does not produce a large response, and there is a great deal of noise in the signal. We need to deal with each of these problems with signal conditioning circuits.

DC BLOCKING AND ZERO-REFERENCING THE SIGNAL

The microphone needs a small amount of power. So it is connected to the +5 V power wire. However, like a voltage divider, some amount of DC voltage is found in its output. This DC component can be blocked by passing the signal through a capacitor. The capacitor should be followed by a high-value resistor that connects the DAQ input to ground. This resistor will assure that no DC voltage can "sit" on the DAQ input wire. Since the DAQ inputs are very high (essentially infinite) impedance, there is no way for a DC potential to be "discharged" to ground. Add a capacitor and resistor as shown in Figure 8.17, Figure 8.13, and Figure 8.14 to block the DC signal and reference the signal to a zero-volt average value. Capacitors and resistors are available on the web and at local electronics stores. We used a Jameco 26956 capacitor and a 270 KQBK-ND Digi-Key resistor. Do not connect any of the other components shown in Figure 8.13, and Figure 8.14 at this point. We will build the circuit a section at a time and measure the signal sequentially throughout this discussion to show the effects of each portion of the circuit.

The BNC-2120 inputs are very high impedance and do not require currents but read voltages only. This makes them vulnerable to noise that stems from electrostatic sources. You should see a buildup in noise as you touch the input wire used to connect to the shielded BNC connector. For this experiment, the signal is conveyed to the BNC connector on the BNC-2120 through a twisted pair wire to a banana jack-to-BNC adapter. Lacking shielding, the static electricity that comes from your body as you touch the input lead will enter the input to the BNC-2120.

FIGURE 8.17
Schematic for DC blocking and zero-referencing

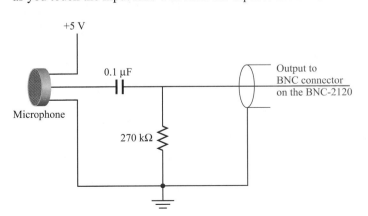

AMPLIFYING THE MICROPHONE SIGNAL

An amplifier will increase the signal voltage level so that it will exceed the noise level. Implement the circuit described in Figure 8.18, Figure 8.19, Figure 8.13, and Figure 8.14 to amplify the signal. Remember to use a twisted pair for the input and output signals to reduce the noise. The resistors and capacitors are available on the web and from local electronics shops. We used Digi-Key 1.0 KQBK-ND, 4.7 KQBK-ND, 47 KQBK-ND, and 100 KQBK-ND resistors and Jameco 11078 capacitor.

As a result of this circuit, we should see that our signal now has a greater response when a sound is applied to the microphone that will allow its signal to rise well above the noise.

This amplifier uses a single NPN transistor in a common emitter configuration that is best suited for voltage amplification. NPN transistors are available on the Web and at local electronics shops. We used a Jameco 26462.

FIGURE 8.18
Microphone amplifier
circuit schematic

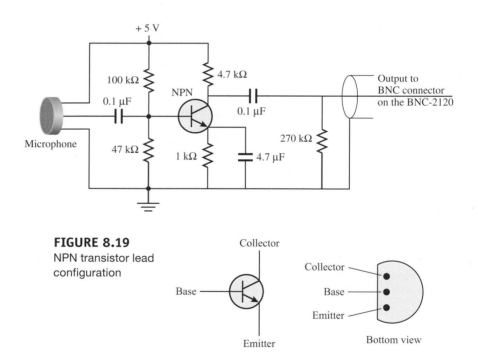

FIGURE 8.19
NPN transistor lead
configuration

APPLICATION IN MUSIC

We are going to apply the fundamentals of high-frequency DAQ to measuring the frequency spectra in music. The audio band of interest is 20 Hz to 20 kHz, commonly accepted as the typical human audio band. The Nyquist theorem recommends that we sample at least 40 kS/s.

Sound is composed of vibrations in air. We will measure these vibrations with a microphone, which essentially measures variations in air pressure. Our measurements can be related to the musical notes in the sound signal because each musical note occurs at a different frequency, as shown in Figure 8.20. If we play only one note, it produces one frequency. Note that half steps in the equal tempered scale are related by the 12th root of 2. A whole note is related by the 12th root of 2 squared. A frequency-to-note converter is located at: http://www.phys.unsw.edu.au/music/note/.

Of course, music is most interesting when several instruments and vocals are played with chords, producing multiple simultaneous frequencies, a sound spectrum, with multiple peaks where loudness or power is represented as amplitude.

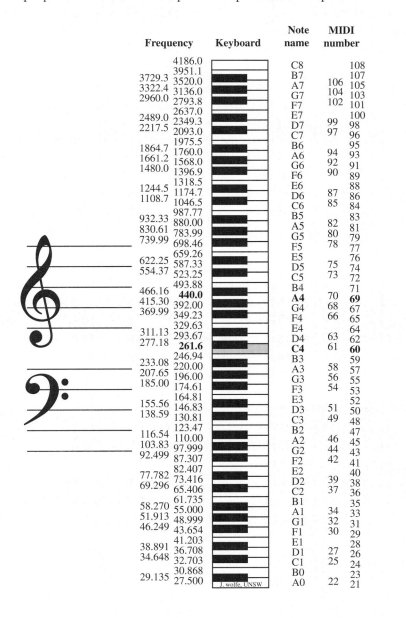

FIGURE 8.20

Musical note frequencies and the piano keyboard

SUMMARY

> Use the DAQmx VIs instead of the DAQ Assistant for high sample rate applications.

> Hardware timing uses the internal clock on the DAQ board.

> The DAQmx Cont Acq&Graph Int-Clk example VI uses the DAQmx VIs and an internal clock.

> Oversampling produces unnecessarily large volumes of data that are difficult to manage, and undersampling biases the data, so determining the correct sampling rate is important.

> The Nyquist theorem states that the highest frequency we can accurately represent is equal to half the sampling rate.

> We can use the Spectral Measurement Express VI to determine the frequencies in a time domain signal.

> Interacting with the user interface (computer monitor) and storage (hard drive) slows execution. So we use multiple loops, one that reads data and others for slower tasks.

> The LabVIEW data-flow paradigm makes it difficult to exchange data with wires between simultaneously executing multiple loops. So we use variables.

> Local and global variables create synchronization and race condition problems when we try to use them to share data between multiple loops, so queues are a better choice.

> The Producer Consumer Design Pattern uses queues to communicate data between multiple loops.

> Many signals require hardware signal conditioning before they are input to a DAQ device.

> The modular approach is recommended for hardware as well as software projects.

EXERCISES

1. Build the Audio Measurement program.

2. Build the microphone signal conditioning circuit in three increments and test with the LabVIEW program after each increment by whistling or speaking loudly into the microphone. Record the signal amplitude and power spectrum result in your notebook by capturing and pasting an image of the front panel window. Comment on the effects on the signal, power spectrum, noise, and amplitudes for each of the three circuits. The first increment is microphone only. The second is microphone and DC blocking and zero-referencing components, and the third is the complete circuit after adding the amplifier.

   a. **First circuit.** Construct the microphone signal directly to the input of the BNC-2120. Connect the red wire from the microphone to a power bus on the protoboard that is hooked to the +5 V supply on the BNC-2120. Connect the black wire to a ground bus on the protoboard that is connected to an analog ground on the BNC-2120. Connect the white wire from the microphone directly to the input of channel zero on the BNC-2120.

   • Observe the quality of the signal compared to the noise level.
   • Observe the DC component of the signal.

   b. **Second Circuit.** Construct the DC blocking and zero-referencing circuit of Figure 8.17.
   • Observe the quality of the signal compared to the noise level.
   • Observe the absence of the DC component of the signal.

   c. **Third Circuit.** Construct the transistor amplifier circuit of Figure 8.18. The transistor leads are shown in Figure 8.18 and Figure 8.19.
   • Observe the quality of the signal compared to the noise level.
   • Observe the DC component of the signal.

3. Disconnect the ground for the microphone. Comment on the effects on the signal, power spectrum, noise, and amplitudes. Then reconnect it to ground through a long wire, forming a large loop, and comment on the effects. Note: These effects may vary widely depending on the EMI levels and the proximity of the ground loop to power.

4. Update your course notebook.

5. Work the example in Chapter 3 of the Getting Started with LabVIEW titled "Analyzing and Saving a Signal."

6. Modify the LabVIEW example DAQmx Continuous Acquire and Graph Internal Clock program to include signal analysis. Add the Spectral Measurement Express VI and a graph so the program displays both time and frequency domain information. Save the program as Sound Measurement. Test and demonstrate it with signals from the BNC-2120 Function Generator.

7. How fast can the PCI-6024E DAQ board sample?

_____

8. Predict the highest frequency that can be measured by the Sound Measurement program without aliasing.

_____

9. Test your prediction with actual measurement. What is the highest sine-wave frequency you can measure from the BNC-2120 Function Generator with the Sound Measurement program without aliasing?

_____

10. Explain any difference between the predicted and actual frequencies.

_____

_____

11. With the BNC-2120 set to generate 100 Hz: What minimum sample rate should you use?

_____

If you sample faster, does the waveform improve?

_____

What is the lowest sample rate that displays a good sine wave?

_____

How does this ratio compare to the Nyquist ratio?

_____

What is the frequency of the peak in the Frequency Domain Graph?

_____

12. Repeat Exercise 11 with the BNC-2120 set to generate 10 kHz and 100 kHz sine waves and report the results in the table below.

	10 kHz	100 kHz
a		
b		
c		
d		
e		

13. If you were unable to display the 100 kHz sine wave very well in Exercise 11 reduce the frequency generated until you get a good display. Report the frequency value below. Explain what limited the ability to obtain a good display.

_____

_____

14. Build a program using the Power of X Function that calculates the 12th root of 2.

15. Using the microphone circuit and the Audio Measurement program. Write the frequencies of the notes played and measured by the Sound Measurement program in the table below.

Note	Frequency	Note	Frequency

16. Use the 12th root of 2 program to calculate the frequencies, and compare with those measured.

17. Add the ability to save the time waveform data to file in the Sound Measurement program. Test the program. Don't run very long to avoid unnecessarily large files.

18. Modify the Sound Measurement program so it uses the State Machine architecture. Test it.

19. Compare the State Machine architecture with the previous program's ability to measure high frequencies.

20. Run the Mechanical Action of Booleans Example VI. Before you press the buttons, list the name of the action and predict what will happen for each action in the space that follows:

_____

_____

_____

_____

_____

21. Explain the difference between your prediction in Exercise 20 and the actual action.

_____

_____

_____

_____

_____

_____

22. Code and run the Parallel Loops program with Execution Highlighting and verify that the second loop runs only once.

23. Code and run the Local Variables program and experiment with time delays to get the second graph to

display every other value (skipping values).

display each value twice (duplicating values).

24. Open the Producer Consumer template and modify it to enqueue random numbers in the producer loop and display them on a waveform chart in the consumer loop.

25. Modify the random number Producer Consumer program from Exercise 25 to enqueue and display temperature data points from the BNC-2120.

26. Modify the random number Producer Consumer program to enqueue sound waveforms and display them.

27. Find the Generate Numbers and Display Numbers examples in LabVIEW. Run them with execution highlighting. Describe how the global variables are used.

_____

_____

28. Write two programs that will run simultaneously. Use a global variable to stop them simultaneously. Each program will contain a While Loop and will display a random number on a chart. Each While Loop will contain a time delay. Use LabVIEW Help to learn how to create and use global variables.

Summary

INTRODUCTION

This chapter summarizes the material presented in the text by developing a versatile and efficient data acquisition system using a LabVIEW state machine. The material presented in this chapter is based on material covered previously, so the chapter is brief. Please refer to previous presentations for more detailed explanations.

OUTLINE

PLAN THE APPLICATION

Our summary application has the states and flow shown in Figure 9.1. The initialization state will open and configure DAQ, open and configure a file, write a header to the file, and initialize shift registers. These resources will be closed in the shutdown state so it is important to always execute the shut-down state before terminating the application. Several events could cause the application to stop: errors, user stop button, or acquiring the amount of data specified by the user. When a stop condition is reached, the application will first go to the shut-down state before termination.

The acquire state passes the DAQ refnum created in the initialize state from a shift register to the DAQmx Read VI. It writes the signal information to a shift register for use by other states. The analyze state computes the FFT of the signal and writes the information to a shift register for use by other states. The display state

FIGURE 9.1 State
diagram for the DAQ
State Machine VI

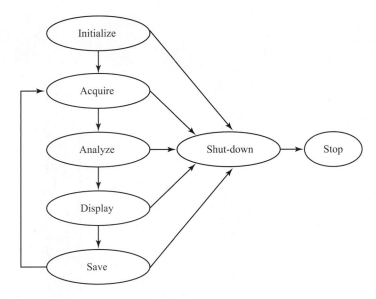

displays both the time and frequency data to front-panel graphs. The save state converts the time signal data to a spreadsheet string and writes the data to a file with the Write to Text File function. The shut-down state closes the DAQ and file resources and closes the references. Errors are passed throughout the states via shift registers. When an error occurs, control passes to the shut-down state where the Simple Error Handler opens the Error dialog box.

FRONT PANEL

The front panel window of the DAQ State Machine, shown in Figure 9.2, gives the user control over several DAQ configuration values. The configuration inputs include the following:

> Number of data sets—a single data set is composed of the samples per channel
> Samples per channel—the number of data points per channel acquired in the acquire state
> Sample rate—the rate used by the DAQ device to acquire data points. This value must conform to the specifications of the DAQ device.
> Physical Channels—the signals to be acquired are physically wired to these channels
> Maximum value—the upper limit of the signal value. This input is used to determine the digitizing resolution. This value should conform to the specifications of the DAQ device.
> Minimum value—the lower limit of the signal value. This input is used to determine the digitizing resolution. This value should conform to the specifications of the DAQ device.
> Input terminal configuration—the way the signal is physically connected to the DAQ system. Common connections are single ended and differential.

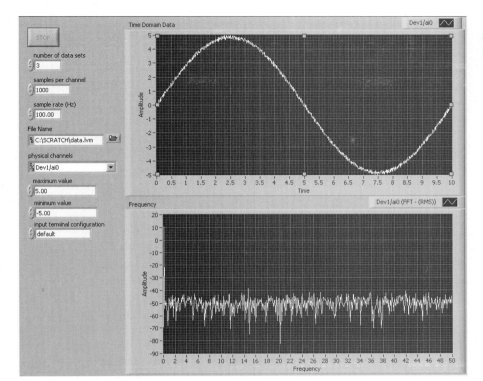

FIGURE 9.2 DAQ State Machine front panel window

INITIALIZATION AND CONFIGURATION

Prior to developing the code on the block diagram window, configure the actual or simulated hardware in MAX, as shown in Figure 9.3.

When the VI runs, the state-transition enumerated constant value of Initialize is passed to the state-transition shift register, and the user-controlled value for number of data sets is passed to a While-loop tunnel.

On every iteration of the While Loop, the value of the Stop control is read to evaluate transition to the shut-down state. The program passes to the shut-down state if Stop is True or if error status is True.

In the first iteration of the While Loop, a DAQ channel is created from the user configuration specifications, as shown in Figure 9.4. This is done with the polymorphic Create Virtual Channel VI that is configured for Analog Input Voltage. This VI creates a reference that is wired to subsequent DAQmx VIs.

The next step is to configure the timing, again from a user input. This is done with the polymorphic DAQmx Timing VI configured to Sample Clock. Then the DAQmx Start Task VI transitions the DAQ task to the running state in preparation for reading the signal. The reference created by the DAQ Create Channel VI is passed

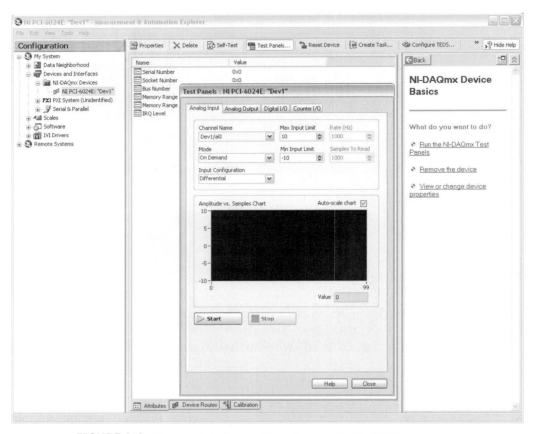

FIGURE 9.3 Simulated device configuration and test in MAX

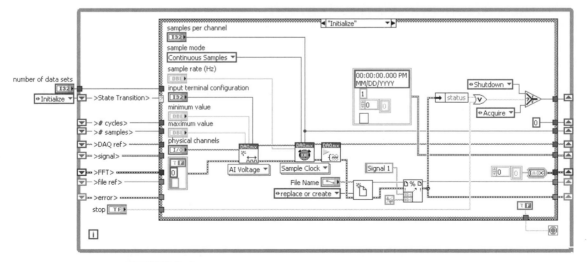

FIGURE 9.4 Initialize state

through the other VIs in the sequence to a shift register. This reference will be closed in the shut-down state.

Next, the block diagram executes the Open/Create/Replace File function; however, this could have been done before initializing the DAQ resource. User-controlled information is input to specify the file name and path. The file operation is set to replace or create. This function creates a reference that is passed to the Format Into File function that executes next to write a header "Signal 1" and insert an end of line character to an ASCII file.

Error data is initialized in this state and errors are passed through the sequence of DAQmx and File VIs and functions. The error data is written to a shift register and also used to send the VI to the Shutdown state if the error status is true.

If there are no errors and if the stop control wasn't selected by the user, the state-transition enumerated constant value of Acquire is passed to the state-transition shift register.

The following constants initialize shift registers:

➤ The signal (waveform data type)
➤ Number of cycles (integer)
➤ FFT (array of doubles)

A Boolean constant value of false is passed to a tunnel and subsequently to the While-loop condition terminal, so the While Loop will continue to iterate, and the program will enter the next state. Only the Shutdown state passes true to the conditional terminal.

ACQUISITION

The DAQ resource reference is passed to the DAQmx Read VI, which is executed in the acquisition state, shown in Figure 9.5. The time-out input is set to -1, so the VI waits indefinitely. The time-out input specifies the amount of time in seconds to wait for samples to become available. If the time elapses, the VI returns an error and returns any samples read before the time-out elapsed. The default value of 10 can cause an error in this VI if the user configuration causes the DAQ task to take longer than 10 s, for example, if the number of samples is over 1000 when the rate is set to 100 Hz (t(s) = 1000/100).

The following data is passed through this state without change:

➤ Number of cycles
➤ Number of samples
➤ FFT
➤ File reference

If the user stop control or the error status is true, execution passes to the shutdown state. Otherwise, the analyze state is executed next.

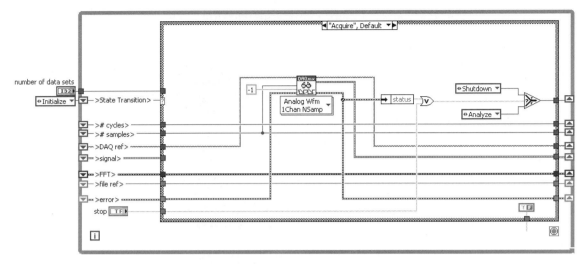

FIGURE 9.5 Acquire state

ANALYSIS

The analysis state, shown in Figure 9.6, computes the FFT of the signal using the Spectral Measurements Express VI, configured as shown in Figure 9.7.

The Express VI requires the input signal to be in the dynamic data type format, so a convert to dynamic data function is used to convert the single waveform data stored in the signal shift register. The analysis state writes the FFT output information to another shift register.

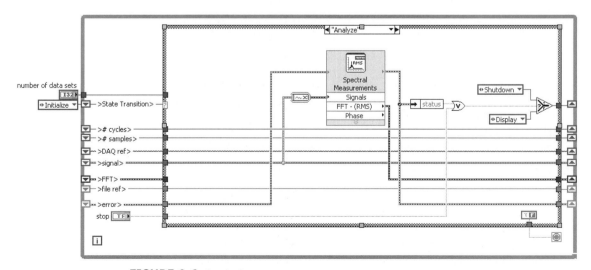

FIGURE 9.6 Analysis state

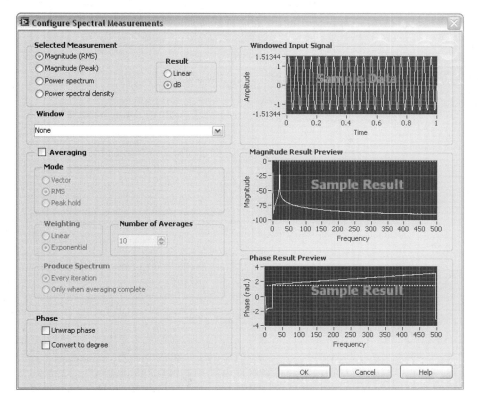

FIGURE 9.7
Spectral
Measurements
Express VI
configuration

The following data is passed through this state without change:

> Number of cycles
> Number of samples
> DAQ reference
> Signal
> File reference

If the user stop control or the error status is true, execution passes to the shutdown state. Otherwise, the display state is executed next.

DISPLAY

Data from the signal and FFT shift registers to waveform graph indicators, as shown in Figure 9.8.

None of the data stored in shift registers is changed in this state.

If the user stop control or the error status is true, execution passes to the shutdown state. Otherwise, the save state is executed next.

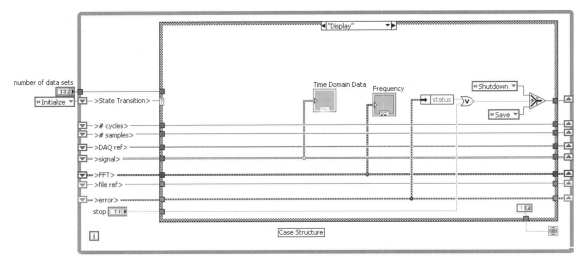

FIGURE 9.8 Display state

SAVING TO FILE

As shown in Figure 9.9, the signal data is read from the shift register, converted to string data type (ASCII format), and written to a file in the save state.

The signal is read from the shift register, and the Y array component is extracted from the waveform data. Next, the Y array is converted to a spreadsheet string (tab

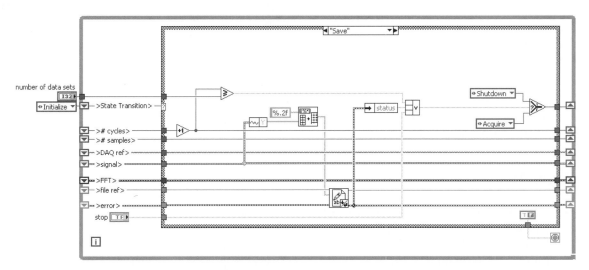

FIGURE 9.9 Save state

delimited) with %.2f format. The resulting string is written to the file using the file reference created in the initialize state and stored in a shift register.

When this state executes, the number of cycles increments. Code in this state compares the number of cycles with the user-specified number of data sets.

The only shift register data that changes in this state is the number of cycles.

If the user Stop control, error status, or number of data sets comparison is True, execution passes to the shut-down state. Otherwise, execution passes to the acquire state.

CLOSING RESOURCES

The shut-down state, shown in Figure 9.10, closes all of the resources that were opened and terminates the application by sending a True value to the While Loop Condition Terminal.

An Empty Task constant is written to the DAQ reference to close the reference. A Not a Refnum constant is used to close the file reference even though it causes a data conversion.

Notice that the program does not have execution control timing. The DAQmx Read execution should provide sufficient relief to the processor for performing other tasks. Also, note that this program acquires a data set and the acquisition state waits while the other states execute. The program architecture can be changed from a state machine to the producer consumer to facilitate more continuous execution of the acquisition process.

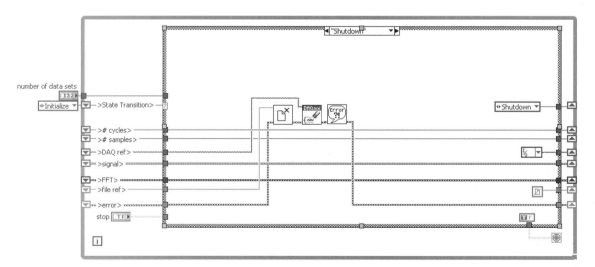

FIGURE 9.10 Shut-down state

EXERCISES

1. Update your course notebook with information from this chapter, including new data types and shortcut keys.

2. Add software control timing to the DAQ State Machine so the application will terminate after a user-specified amount of time or if the number of samples per channel has been acquired. Modify the State Machine so the user can choose whether to use run time or number of samples per channel as the termination criteria.

3. Change the time-out on the DAQmx Read VI from −1 to 10 and try to collect 1000 samples per channel. What happens? Why?

   _____

   _____

   _____

4. Change the front-panel maximum and minimum controls so they will show only the values allowed for your hardware. For example, the PCI-6024E board input ranges are −10 to 10 V, −5 to 5 V, −500 to 500 mV, −50 to 50 mV.

5. Use the LabVIEW Help to learn about the LabVIEW project. Open a project for the DAQ State Machine VI. Test the VI in the project. Why might you want to use a project for this and other applications?

   _____

   _____

   _____

6. Modify the DAQ State Machine to acquire data from an instrument, using the Instrument Assistant Express VI instead of the DAQmx VIs.

7. Modify the DAQ State Machine to use the Producer Consumer architecture with the acquire code in the producer loop and the remaining code in the consumer loop.

8. After modifying the DAQ State Machine to use the Producer Consumer architecture (Exercise 7), add code that will check the number of elements left in the queue. Empty the queue before terminating the application.

9. Benchmark this application to determine the amount of time to execute each state and the entire application. Study performance benchmarking in LabVIEW Help and on ni.com. Execute the Get Time and Date in Seconds function before the State Machine starts, and subtract the start time from the time in each state by adding a Get Time function in each state. Write the times to a file after the state machine terminates. Modify the application to run for 100 cycles through the State Machine. Look at the variability of time over the 100 cycles.

10. Determine the execution time for the Get Time and Date in Seconds function, and subtract the time from the results in the previous benchmarking exercise.

11. Do some research on the difference between the MSWindows and real-time operating systems. What would happen to the variability in time required to execute the DAQ State Machine in MSWindows versus executing it in a real-time OS?

    _____

    _____

    _____

    _____

12. Divide the DAQ State Machine into two VIs running on the same computer, one that acquires data and one that receives and processes the data. Put the two VIs in a project. Use a shared variable to communicate between VIs.

13. Divide the DAQ State Machine into two VIs running on different computers and sharing data over an Ethernet connection. One computer will acquire and transmit the data, and the other will receive and process the data. Put the two VIs in a project. Use a network-published shared variable to communicate between VIs.

14. Divide the DAQ State Machine into two VIs running on different computers and sharing data over an Ethernet connection. One computer will acquire and transmit the data and the other will receive and process the data. Put the two VIs in a project. Use TCP to communicate between VIs.

REFERENCES

Austin, M., and D. Chancogne. *Introduction to Engineering Programming: C, MATLAB, JAVA.* John Wiley and Sons, 1999.

Bishop, R. H. *Learning with LabVIEW Express.* Pearson Prentice Hall, 2004.

Deitel, H. M., and P. J. Deitel. *C++ How to Program.* 4th ed. Prentice Hall, 2003.

Holloway, J. P. *Introduction of Engineering Programming Solving Problems with Algorithms.* John Wiley and Sons, 2004.

Ifeachor, E. C., and B. W. Jervis. *Digital Signal Processing—A Practical Approach.* 2nd ed. Prentice Hall, 2002.

Ingber, D. M., and D. M. Etter. *Engineering Problem Solving with C++.* Prentice Hall, 2003.

Patterson, D. A., and J. L. Hennesy. *Computer Organization and Design—The Hardware/Software Interface.* 3rd ed. Morgan Kaufman, 2003.

Ritter, D. *LabVIEW GUI: Essential Techniques.* McGraw-Hill, 2002.

Travis, J. and J. King *LabVIEW for Everyone.* 3rd ed. Prentice Hall, 2002.

Figures:

Chapter 1

1.1: Robert H. King
1.2: © 2008 National Instruments Corporation
1.3: © 2003 National Instruments Corporation
1.4: © 2003 National Instruments Corporation
1.5: © 2008 National Instruments Corporation
1.6: © 2008 National Instruments Corporation
1.7: © 2007 National Instruments Corporation

Chapter 2

2.23: Copyright © 1999 Austin. Reprinted with permission of John Wiley & Sons, Inc.

Chapter 3

3.3: © National Instruments Corporation
3.4: © 2008 National Instruments Corporation
3.14: © 2007 National Instruments Corporation
3.15: © 2003 National Instruments Corporation
3.16: © 2003 National Instruments Corporation
3.21: © 2008 National Instruments Corporation

Chapter 6

6.8: © 2007 National Instruments Corporation
6.9: © 2007 National Instruments Corporation
6.10: © 2007 National Instruments Corporation

Chapter 7

7.19: © 2007 National Instruments Corporation
7.25: © 2008 National Instruments Corporation
7.26: © 2007 National Instruments Corporation
7.27: © 2007 National Instruments Corporation

Chapter 8

8.1: © 2007 National Instruments Corporation
8.2: © 2007 National Instruments Corporation
8.14: Robert H. King
8.20: Courtesy of Music Acoustics, www.phys.unsw.edu.au/music

Tables:

Chapter 1

1.1: © 2003 National Instruments Corporation

Chapter 2

2.1: © 2007 National Instruments Corporation
2.2: © 2007 National Instruments Corporation
2.3: © 2007 National Instruments Corporation
2.4: © 2007 National Instruments Corporation

Chapter 3

3.1: © 2000 National Instruments Corporation

Chapter 7

7.1: © 2007 National Instruments Corporation
7.2: © 2007 National Instruments Corporation
7.3: © 2008 National Instruments Corporation
7.4: © 2008 National Instruments Corporation
7.5: © 2008 National Instruments Corporation

Exercises:

Chapter 4

Copyright © 1999 Austin. Reprinted with permission of John Wiley & Sons, Inc.

A

Abort, 35, 87, 111, 168
acceleration, 192
accuracy, 5, 7, 62
AIGND, 60
algebra, 9
algorithms, 9, 74–77, 100, 139
alignment, 31, 34
American Standard Code for Information Interchange (ASCII),
164, 170, 180–182, 217, 220
analog input, 62
analog input sense (AISENSE), 60
analog to digital converter (ADC)
digitizing resolution and, 57–59
selecting DAQ device for, 62
"Application Design Patterns: State
Machines" (Hogan), 109
arithmetic, 35–36
arrays
advantages of, 123
Block Diagram and, 123–127
Boolean, 123–124, 147, 154
Build function, 133–134
clusters and, 123–124, 154–155
computer storage and, 123
as data structure scheme, 123
data type coercion and, 122, 139
defining, 123–126
dimensions and, 123, 126
disk streaming and, 170–171
elements and, 123, 132–134
file tasks and, 163–165
For Loop and, 126, 133
Formula Node and, 127–128, 133
Front Panel and, 123–126
function commands for, 132–134
graphs/charts and, 129–132
histograms and, 142–144
identifiers and, 123

indexing and, 123, 128–129,
133–135, 139, 141–142,
147–150, 154–155
loop tunnels and, 126–127
matrices and, 149–153
memory and, 123–127, 133
multidimensional, 134–139
noise and, 139–142
number–to–string conversion and,
164–165
numeric, 123
path, 123–124
picture indicator and, 154–155
refnum, 124
Replace Array Subset, 133
saving and, 220–221
shift registers and, 147
sinusoid signal generation and,
127–128
statistical analysis and, 139–144
string, 123–124
subarrays, 133–134
topographic maps and, 123
waveform, 123
Austin, M., 15

B

Banana Jack–to–BNC adapter,
205, 207
bias, 5, 209
Block Diagram
arrays and, 123–127
Case Structure and, 71
clusters and, 154–155
configuration and, 215–217
data types and, 82
debugging and, 38
developing code on, 21–25
Execution Highlighting and, 38
expandable nodes and, 118
For Loop and, 126

Formula Node and, 127–128, 133
icons and, 118
initialization and, 215–217
Instrument I/O Assistant Express
VI and, 184
local/global variables and, 199–200
loop tunnels and, 126–127
modular programming and, 115
program structure and, 96–99
software design fundamentals
and, 71
Spectral Measurements Express VI
and, 195
statistical analysis and, 141–142
style and, 111–112
units and, 40–44
BNC–2120, 49–51
Banana–Jack–to–BNC adapter
and, 205, 207
connection diagram and, 63–64
DAQ Assistant and, 54–66
function generator, 64–65
impedance inputs and, 207
Measurement & Automation
Explorer (MAX) Program
and, 51–53, 64
microphone circuit and, 203–207
protoboards and, 203–204
quadrature encoder and, 65–66
terminal configuration and, 59–61
user preferences and, 72
Boolean Control, 71
Boolean data types
arrays and, 123–124, 147, 154
Case Structure and, 80–82
high–frequency sampling and,
200–201
logical behaviors for, 201
polymorphism and, 101–102
switch/latch logic and, 200–201
While Loop and, 83–84